ACS SYMPOSIUM SERIES 631

# Chemical Markers for Processed and Stored Foods

**Tung-Ching Lee,** EDITOR
*Rutgers, The State University of New Jersey*

**Hie-Joon Kim,** EDITOR
*U.S. Army Natick Research,
Development and Engineering Center*

Developed from a symposium sponsored
by the Division of Agricultural and Food Chemistry
at the 210th National Meeting
of the American Chemical Society,
Chicago, Illinois,
August 20–24, 1995

American Chemical Society, Washington, DC

Sep/Ae
chem

**Library of Congress Cataloging-in-Publication Data**

Chemical markers for processed and stored foods / Tung-Ching Lee, editor, Hie-Joon Kim, editor.

　　p.　　cm.—(ACS symposium series, ISSN 0097–6156; 631)

"Developed from a symposium sponsored by the Division of Agricultural and Food Chemistry at the 210th National Meeting of the American Chemical Society, Chicago, Illinois, August 20–24, 1995."

Includes bibliographical references and indexes.

ISBN 0–8412–3404–3

1. Food industry and trade—Quality control—Congresses.
2. Food—Analysis—Congresses.

　　I. Lee, Tung-Ching, 1941–　. II. Kim, Hie-Joon, 1947–　.
III. American Chemical Society. Division of Agricultural and Food Chemistry. IV. American Chemical Society. Meeting (210th: 1995: Chicago, Ill.) V. Series.

TP372.5.C484　1996
664′.07—dc20
　　　　　　　　　　　　　　　　　　　　96–19198
　　　　　　　　　　　　　　　　　　　　　　CIP

This book is printed on acid-free, recycled paper.

# Foreword

THE ACS SYMPOSIUM SERIES was first published in 1974 to provide a mechanism for publishing symposia quickly in book form. The purpose of this series is to publish comprehensive books developed from symposia, which are usually "snapshots in time" of the current research being done on a topic, plus some review material on the topic. For this reason, it is necessary that the papers be published as quickly as possible.

Before a symposium-based book is put under contract, the proposed table of contents is reviewed for appropriateness to the topic and for comprehensiveness of the collection. Some papers are excluded at this point, and others are added to round out the scope of the volume. In addition, a draft of each paper is peer-reviewed prior to final acceptance or rejection. This anonymous review process is supervised by the organizer(s) of the symposium, who become the editor(s) of the book. The authors then revise their papers according to the recommendations of both the reviewers and the editors, prepare camera-ready copy, and submit the final papers to the editors, who check that all necessary revisions have been made.

As a rule, only original research papers and original review papers are included in the volumes. Verbatim reproductions of previously published papers are not accepted.

ACS BOOKS DEPARTMENT

# Contents

CHEMICAL MARKERS OF STORAGE EFFECTS

# Preface

WHO NEEDS CHEMICAL MARKERS? Chemical markers are compounds that can be used to evaluate quality changes either during processing or during subsequent storage of foods at various conditions. By virtue of their quantitative, predictive, and mimicking features, chemical markers are useful for food scientists and technologists who need to maximize the initial product quality and minimize the deteriorative changes in food during processing and storage. Information about the reaction mechanisms as well as new sensor technologies and other instrumental and analytical methodologies make the use of chemical markers a practical tool in many food applications in-line, on-line, and off-line.

Chemistry is involved in every aspect of the food cycle[1]. Of the many chemical reactions in the food cycle, the reactions taking place during processing and storage are particularly important to food chemists. Some reactions, such as the formation of flavor compounds in baking, are desirable. Other reactions, such as lipid oxidation in meats, lead to rancidity and are undesirable. To complicate matters, the desirable and undesirable reactions often represent different stages of the same series of reactions. For example, the Maillard reaction is important for flavor and color generation. However, advanced Maillard reactions could also lead to lysine loss and melanoidin formation or protein cross-linking, as well as other reactions that could cause detrimental nutritional, physiological, and toxicological consequences in certain foods[2]. Understanding the chemical reactions and being able to control the reaction pathway or the extent of the reaction could be critical for optimizing the product quality upon processing and for minimizing detrimental changes during processing and storage. When judiciously used, chemical markers could be useful tools for food scientists because of the following capabilities.

• *Chemical markers provide quantitative information.* The ultimate test of a good food product is acceptance by the consumer. This acceptance can be measured by a test panel and quantified to some extent. However, such panel results are often subjective, and objective quantitation using chemical markers is more often desirable. Quantitation by color machine version (Chapter 23) is an example. Depending on the reaction rate constant, quantitative results can be obtained at thermal processing temperature (Chapter 6), at ambient storage temperatures (Chapter 9), and at frozen storage temperatures (Chapter 15).

• *Chemical markers have predictive capabilities.* Quantitative results derived from chemical marker measurements can be useful for predicting what subsequent reactions will follow once the predictive model is established. Some interesting examples in terms of predicting remaining shelf-life are provided in Chapter 10. Moreover, detection of early reaction products or intermediate compounds enables us to predict the extent of subsequent reactions as demonstrated in Chapter 4.

• *Chemical markers can mimic other processes.* Chemical reactions taking place in foods can be used to mimic an entirely different process taking place in the vicinity of the site where the chemical measurements are made. An interesting example is the use of chemical markers to determine lethality within a food particulate where direct temperature measurement is not practical (Chapter 6). It appears that chemical reactions can mimic bacterial destruction, and are potentially useful time–temperature integrators in the continuous thermal processing of foods.

In conjunction with the future development of better quantitive chemical markers, the most significant imminent trend in food quality control is the development and application of new sensor technologies and other instrumental methods for the in-line, on-line, and off-line quantitative determination of these chemical markers. This trend is well-illustrated in many chapters in this book (e.g., Chapters 4–7, 9, 12, 15, 16, 18, 19, 21–23). Furthermore, accurate kinetic models are needed to be able to predict remaining shelf-life as well as to optimize product quality that depends on the intricate interplay of various chemical reactions under various processing conditions.

Today, people are concerned about the quality of food they consume. The questions are, "What are the important food quality attributes and how can we determine these attributes in-line, on-line, and off-line?" We believe this book is timely and informative in addressing these issues. We hope this book will have wide appeal and will be useful to people working in academia, the food industry, and in governmental agencies dealing with the food manufacturing and service industries, as well as those interested in food quality and its evaluation and controls.

**Literature Cited**

[1]*Chemistry of the Food Cycle: State of the Art;* Taub, I. A.; Karel, M., Eds. Reprinted from *J. Chem. Educ.* **1984**, 61(4), 271–367.

[2]Chichester, C. O.; Lee, Tung-Ching. "Effect of Food Processing in the Formation and Destruction of Toxic Constituents of Food." in *Impact of*

*Toxicology in Food Processing;* AVI Publishing Company: Westport, Ct, 1981, pp 35–56.

TUNG-CHING LEE
Department of Food Science and
Center for Advanced Food Technology
Cook College
Rutgers, The State University of New Jersey
P.O. Box 231
New Brunswick, NJ 08903–0231

HIE-JOON KIM
U.S. Army Natick Research
Development and Engineering Center
Kansas Street
Natick, MA 01760–5018

March 28, 1996

# CHEMICAL MARKERS
## OF PROCESSING EFFECTS

# Chapter 1

# Analysis of Nonvolatile Reaction Products of Aqueous Maillard Model Systems

## The Effect of Time on the Profile of Reaction Products

**Jennifer M. Ames, Richard G. Bailey, Simona M. Monti[1], and Cheryl A. Bunn**

**Department of Food Science and Technology, University of Reading, Whiteknights, Reading RG6 6AP, United Kingdom**

Aqueous one molal solutions of xylose and lysine monohydrochloride were refluxed for up to 120 min with the pH maintained at 5. Analysis of reaction mixtures by diode array HPLC revealed changes in the chromatograms with time of heating. Reaction products could be divided into four groups: unretained material, resolved compounds and two types of unresolved substances. Storage of the model system, heated for 15 min, at 40 °C for up to three weeks resulted in changes to the HPLC chromatograms which were different from those observed on prolonged heating. Two peaks became prominent on the chromatograms of the heated samples and one peak became prominent on the chromatograms of the stored samples. The progressive development of a complex of partially resolved peaks on storage at 40°C provided another means of distinguishing between the heated and stored samples. The different chromatographic behaviours observed in samples subjected to prolonged heating or storage suggest that the presence of certain marker compounds in the system may provide information regarding sample history.

The Maillard reaction is chiefly responsible for the desirable colors and flavors that occur when many foods are heated. A substantial amount of data has been obtained relating to the volatile aroma compounds produced during the reaction (*e.g. 1-3*), and this is due largely to the success of gas chromatography-mass spectrometry as a technique for the separation and identification of volatile compounds. Information relating to the factors affecting the formation of these aroma compounds is also available (*e.g. 4-7*).

Very much less data is available relating to the colored compounds that are formed during the Maillard reaction. Such compounds may be divided into the low

[1]Permanent address: Università degli Studi di Naoli Federico II, Facoltà di Agraria, Dipartimento di Scienza degli Alimenti, Parco Gussone, 80055 Portici, Italy

0097–6156/96/0631–0002$15.00/0
© 1996 American Chemical Society

molecular weight reaction products, that comprise two to four linked rings, and the melanoidins which are colored, nitrogenous, polymeric materials with masses of up to several thousand daltons (*8*). A range of low molecular weight colored compounds has been isolated from Maillard model systems over the last twenty years (*8-10*), and the structures of several of them have been characterized in our laboratory (e.g. *10, 11*). Most of the reports dealing with the low molecular weight colored compounds formed during the Maillard reaction focus on the elucidation of the structures of the reaction products rather than the parameters that may affect the reaction product profile.

The work described here is part of a continuing project aimed at investigating the formation of non-volatile compounds, both colorless and colored, in Maillard model systems (*9, 10*). The focus of this report is the effect of time of heating and subsequent storage on the profile of reaction products formed from an aqueous xylose-lysine hydrochloride model system.

**Experimental**

Aqueous xylose-lysine monohydrochoride model systems (one molal with respect to each reactant) were refluxed for up to 120 min. The pH was maintained at 5 throughout heating by the periodic addition of 3 M sodium hydroxide. Samples were cooled in ice before analysis by HPLC. Aliquots of the model systems that had been refluxed for 15 min were stored at 40 °C for up to three weeks and analyzed at weekly intervals.

Diode array HPLC was performed using a Hewlett-Packard (Bracknell, UK) 1050 quaternary pump, autosampler and diode array detector and data were analyzed using HP Chemstation software. Separations were performed using an ODS2 column and a linear water/methanol gradient. Full experimental details are given in Bailey *et al.* (*10*).

**Results and Discussion**

The methods developed for this work allowed the direct analysis of the total reaction products by diode array HPLC. This was important since, during the development of procedures, it became apparent that the chromatograms obtained on injecting fresh model systems after 6 and 15 min heating varied when injections were repeated 70 min later (the HPLC run time) with the sample stored in the autosampler at room temperature. This suggested that any sample pre-treatment prior to HPLC would be likely to result in a modified reaction products profile. Chromatograms were obtained at 254 and 280 nm (for detection of colorless compounds), 360 nm (for detection of pale yellow components) and 460 nm (for colored material).

The model system changed color from pale yellow through orange to dark brown on heating. For the heated systems, the sample components separated by HPLC could be placed into four groups (*10*) as shown in Fig. 1. The first group comprised the unretained peaks that eluted in the void volume of the column and possessed retention times between 2 and 3 min. The diode array spectra of these peaks on the 460 nm chromatogram showed broad absorption bands in the region 320-360 nm with tailing into the visible region. Such spectra are characteristic of brown compounds (*12*) and may indicate the presence of melanoidins. Resolved peaks with retention times between 4 and 25 min made up the second group, and consisted of both colorless and colored compounds. The fact that they were resolved or partially resolved by the column indicates that they were of low or intermediate molecular weight (*13*). Components that ran as a convex broad band described the third type of behaviour, and spectra from the band corresponded to those of a brown compound. This type of band indicated the presence of brown polymeric material (melanoidin). The fourth group of components consisted of unresolved material that ran as a tailing broad band, observed on the 254, 280 and 360 (but not the 460) nm chromatograms as a raised baseline at *ca.* 3 min which tailed to the horizontal. This band also indicates polymer which is either colorless or pale yellow. Thus, it appears that two types of polymer are formed in the heated systems and that they may be distinguished by both their absorption spectra and chromatographic behaviour, both features suggesting a different chemical nature and/or a different range of relative molecular mass.

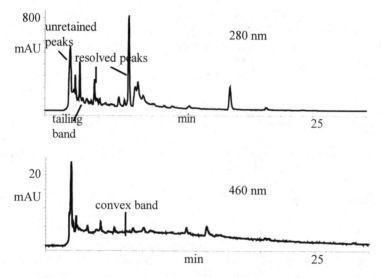

Fig. 1 280 and 460 nm HPLC traces for the model system heated for 15 min, showing the four types of reaction products

    The systems stored at 40 °C behaved untypically. After storage for one week, a group of partially resolved peaks, with retention times in the range 3 to 10 min, appeared on the chromatograms at all wavelengths. This group of peaks became progressively less well resolved with increasing storage time and, by three weeks, it is probably best described as a localised unresolved broad band. An explanation of this behaviour is that the substances responsible consisted of a mixture of oligomers whose molecular mass increased on storage, to give an unresolved mixture, but with a narrow distribution of molecular mass compared with the broad bands observed on prolonged heating. After storage for the three weeks, the column could not be easily regenerated even after extensive washing with 2-methylethanol. This behaviour was not observed for any of the samples obtained by refluxing for times of up to 120 min without storage (*10*). Analysis of the diode array spectra from the complex partially resolved mixture, after storage for two weeks, suggested that the mixture consisted of a range of compounds, differing in colour. The first few peaks were probably primarily yellow in colour, the remaining peaks being brown. After storage for three weeks, the chromatographic behaviour of the sample components and the nature of the diode array spectra indicated the presence of polymeric material. The 460 nm chromatogram of the sample stored for three weeks is shown in Fig. 2 and may be compared with the 460 nm chromatogram of the same sample before storage shown in Fig. 1. It seems that the oligomeric and polymeric materials formed on storage at 40 °C are able to coat the stationary phase of the HPLC column and prevent it from operating effectively. A modified method of analysis is required for these samples.

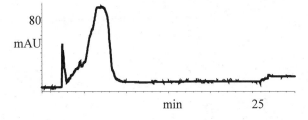

Fig. 2 460 nm HPLC trace for the model system refluxed for 15 min followed by storage at 40 °C for three weeks

The chromatograms obtained at different heating and storage times are discussed below in relation to the groups of components described above.

**Heated Model Systems.** Heated model systems were analyzed by HPLC at zero min (the time at which the sugar and amino acid had dissolved), 6 min (the heating time required before refluxing began) and after heating for 15, 30, 60, 90 and 120 min. The chromatograms obtained at 280 nm after 15, 30, 60 and 120 min are shown in Fig. 3. The groups of reaction products detected at the four monitored wavelengths with reaction time are summarized in Table 1.

Only unretained components appeared in the chromatograms of the model system at zero time and were the first indication that some reaction had occurred. Resolved peaks were observed in the 6 min sample. Unretained peaks at 254, 280 and 360 nm and the tailing broad band were also present after 6 min.

The 15 min sample showed the greatest range of resolved components with peaks being observed at all wavelengths. The chromatogram at 280 nm contained the greatest number of peaks; about 20 main peaks being detected. Seven main peaks, all of which were due to coloured components, were observed on the 460 nm chromatogram. About 20 diode array spectra were obtained for peaks from the 360 nm trace; some being very small. They showed absorption in the region 300-360 nm and it

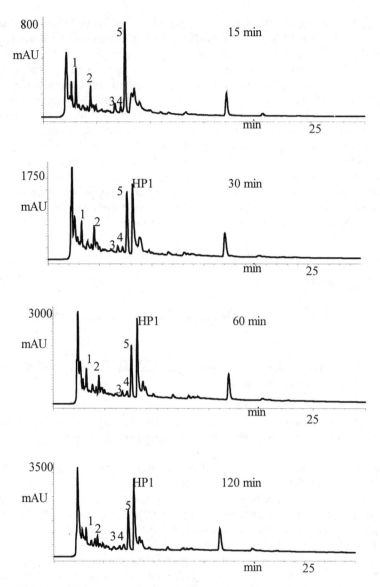

Fig. 3 280 nm HPLC traces for the model system refluxed for up to 120 min

is suggested that they were due to pale yellow compounds. Both types of unresolved broad band were observed for the 15 min sample.

    After 30 min heating, unresolved material was present at all wavelengths in all samples. At 360 nm, resolved peaks were detected and some increased in size with further heating. Resolved peaks were observed in all samples at 460 nm, but the

Table 1 Wavelengths (nm) of detection of reaction product groups with time of heating

| Heating time (min) | Unretained peaks | Resolved peaks | Tailing broad band | Convex broad band |
|---|---|---|---|---|
| 0 | 254, 280 | - | - | - |
| 6 | 254, 280, 360 | 254, 280, 360, 460 | 254, 280, 360 | - |
| 15 | 254, 280, 360, 460 | 254, 280, 360, 460 | 254, 280, 360 | 460 |
| ≥ 30 | 254, 280, 360, 460 | 254, 280, 360, 460 | 254, 280, 360 | 460 |

chromatograms at this wavelength became increasingly dominated by the convex broad band with further heating, and this was due to melanoidin. The height of the convex band increased and its maximum height shifted to longer retention times with increased reaction time (*10*).

Relatively few studies have monitored the effect of heating time on the profile of reaction products in Maillard model systems at temperatures relevant to food processing. As far as studies relating to colour are concerned, it is only reported that the intensity of brown colour, determined by absorbance measurements at a single wavelength, increases with time (e.g., *15*). More studies have been published relating to the volatile reaction products and it appears that total yields of volatiles and yields of some individual classes increase up to a certain point before decreasing (e.g *16, 17*). In contrast, other chemical classes appear after a certain heating time and are formed in increasing amounts with progressive heating (*17*). For example, in a study of an aqueous xylose-lysine monohydrochloride system in which the pH was uncontrolled throughout heating for 4 h, yields of total volatiles and most non-nitrogen containing classes increased up to 2 or 3 h heating, followed by a small decrease in yield. In contrast, representatives of most nitrogen-containing classes, including pyrroles, were only identified after heating for a specific time but then increased in quantity with progressive heating (*17*). One goal of the work reported here is to obtain similar information relating to the formation of specific compounds or classes of compound, which are non-volatile, with time of heating.

**Stored Model Systems.**   The 15 min sample was stored at 40 °C for up to three weeks to monitor the effects of storage on the reaction products profile. Changes observed after storage for one week included the disappearance of a band of poorly resolved components eluting between 9 and 12 min. After two weeks, a complex of partially resolved peaks was observed at all monitored wavelengths which became progressively less well resolved with storage time. After three weeks, this was accompanied by a rapid loss of resolving power of the column, probably caused by polymer contamination. Samples could not be analysed after three weeks storage without sacrificing the HPLC column, since the column could not be regenerated. Changes observed for the 15 min sample at 280 nm are shown in Fig. 4.

Before storage

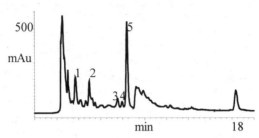

One week storage

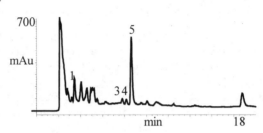

Two weeks                                                                    storage

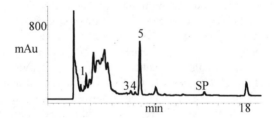

Three weeks storage

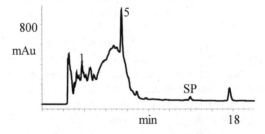

Fig. 4 280 nm HPLC traces for the model system refluxed for 15 min followed by storage at 40 °C for up to three weeks

**Selected peak monitoring with time of heating or storage.**     The peak areas of selected peaks observed on the 280 nm chromatogram for the 15 min sample were monitored over time of refluxing or storage at 40 °C. Two peaks (Heating Peaks, HP1 and HP2; Figs. 3 and 5) became prominent with progressive heating and one peak (Storage Peak, SP; Fig. 4) became prominent with time, and these were also monitored. Peaks in different chromatograms were confirmed as being due to the same compound by reference to both HPLC retention times and diode array spectra. The peaks chosen for monitoring are given in Table 2. A comparison of the diode array spectra of peaks 1-5 with a computer library of standard compounds showed that all of them, apart from peak 3, possessed furanone-like spectra (*14*). Peak 3 could not be assigned to any chemical class. The storage peak, which became prominent in the sample after storage, possessed a furan-like spectrum, but with an additional $\lambda_{max}$ at 335 nm. It is most likely to be due to a derivative of furfural. The two peaks that became prominent with progressive heating could not be assigned to any chemical class. The fact that peaks 1, 2, 4 and 5 possess almost identical UV spectra suggests that they possess similar structures. All the peaks, apart from HP2, were monitored at 280 nm. HP2 was monitored at 360 nm (see Fig. 5).

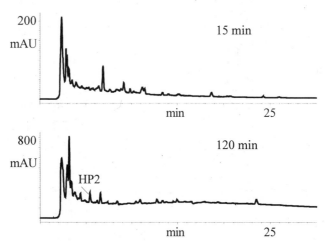

Fig. 5 360 nm HPLC traces for the model system refluxed for 15 and 120 min

Fig. 6 shows a representative plot of changes in peak area with time of heating for all the monitored peaks. Amounts of peaks 1-5 increased steadily up to 60 or 90 minutes refluxing followed by a slight decrease, possibly due to participation in further reactions such as those leading to polymers. In contrast, HP1 was present at only low levels after 15 min heating but, after 30 min, it was the largest peak on the 280 nm chromatogram. Levels of this peak generally increased with heating time. Although amounts did seem to decrease after 90 min, HP1 was still a dominant feature of the chromatogram and levels increased again by 120 min. Its presence at a high level provided a means of distinguishing samples that had been refluxed for 30 min and longer from samples that had been refluxed for 15 min followed by storage. HP2

Table 2 Retention times and diode-array spectral data for peaks monitored over time of heating and storage

| Peak | Retention time (min) | $\lambda_{max}$ (nm) | Tentative assignment |
|---|---|---|---|
| 1 | 3.22 | 270 sh, 300 | furanone |
| 2 | 4.62 | 270 sh, 300 | furanone |
| 3 | 6.87 | 345, 270 | unknown |
| 4 | 7.40 | 270 sh, 300 | furanone |
| 5 | 7.80 | 270 sh, 300 | furanone |
| Heating peak 1 | 5.46 | 280 | unknown |
| Heating peak 2 | 8.12 | 325 | unknown |
| Storage peak | 13.93 | 280, 230, 335 | furan derivative |

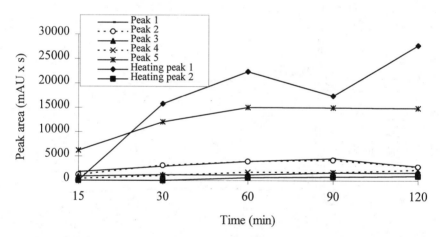

Fig. 6 A representative plot showing changes in peak areas with time of heating for selected peaks

was monitored at 360 nm. It was present after 15 and 30 min heating but was only clearly visible after 60 min and it became progressively larger in the 90 min and 120 min samples. Of all the peaks monitored in the heated samples, this peak was present at the lowest levels. It may also have potential for use as an indicator of samples that have been heated for an extended length of time.

A representative plot of changes in peak area with time of storage of the 15 min sample at 40 °C is shown in Fig. 7. Peaks 1 and 3-5 are the same peaks on Figs. 4 and 6. On storage, peaks 1 and 3 decreased steadily while peaks 4 and 5 showed a small increase after one week, and then decreased. Levels of peak 2 could not be

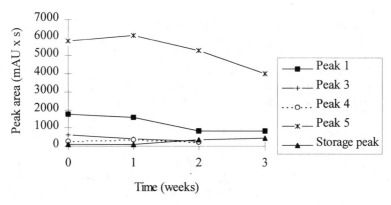

Fig. 7 A representative plot showing changes in peak areas with time of storage at 40 °C for selected peaks

monitored with time of storage since it was not sufficiently well resolved. These behaviours are different to those observed for the same peaks with prolonged heating.

The storage peak was just discernible on the 280 nm chromatogram before storage. It was a small peak but it increased in size after 2 and 3 weeks storage. Its presence was helpful in distinguishing between 15 min samples stored for two weeks or longer and those that had been refluxed for 30 min or longer but not stored. However, the presence of a complex partially resolved mixture of components which eluted between 3 and 10 min and which increased in size with time of storage at all four monitored wavelengths (Fig. 4) was the best way of distinguishing between stored and unstored samples.

Further work is required to establish whether samples that have been subjected to different combinations of reflux and storage conditions may be distinguished from each other.

**Conclusions**

A highly complex mixture of non-volatile compounds is formed in Maillard model systems comprising both colorless and colored material. The reaction between xylose and lysine monohydrochloride in water begins as soon as the reactants are mixed at room temperature. Color develops quickly on heating and before the system has begun to reflux. Reaction products could be classified into four groups, according to their chromatographic behaviour; low or intermediate molecular weight compounds and two types of polymeric material were present. HPLC with diode-array detection allows the change in reaction product profile to be monitored with time of heating.

After storage of the 15 min sample at 40 °C, a complex of partially resolved components develop in the system that eventually cause deterioration of the HPLC column. This behaviour was not observed for any of the samples heated for up to 120 min that had not been stored. It appears that changes occur in the stored sample that are different from changes that occur to the sample on extended heating.

Additionally, two peaks became prominent on the chromatograms of samples heated for 30 min and longer, and one peak, combined with the development of a complex of partially resolved material, helped to distinguish the samples that had been stored for more than one week. These chromatographic features suggest that marker compounds may be formed during the Maillard reaction under different sets of conditions that would provide information concerning the history of the sample.

## Acknowledgments

J M Ames and R G Bailey thank the BBSRC, UK for financial support and S M Monti thanks The University óf Naples "Federico II", Italy for a scholarship in support of this work. The authors are very grateful to Professor J Mann, Department of Chemistry, The University of Reading, for many helpful discussions and to Professor H E Nursten, Department of Food Science and Technology, The University of Reading, for helpful suggestions. The Royal Society and The University of Reading (The University of Reading Research Endowment Fund) are thanked for travel grants (to JMA).

## Literature Cited

1.	Nursten, H. E. *Food Chem.* **1980,** *6,* 263-277.
2.	*Thermal Generation of Aromas;* Parliment, T. H.; McCorrin, R. J. and Ho, C.-T., Eds.; ACS Symp. Ser. 409, American Chemical Society, Washington D. C., 1989.
3.	*Thermally Generated Flavors. Maillard, Microwave and Extrusion Processes;* Parliment, T. H.; Morello, M. J. and McGorrin, R. J., Eds.; ACS Symp. Ser. 543, American Chemical Society, Washington D. C., 1994.
4.	Shu, C.-K. and Ho, C.-T. In *Thermal Generation of Aromas;* Parliment, T. H.; McCorrin, R. J. and Ho, C.-T., Eds.; ACS Symp. Ser. 409, American Chemical Society, Washington D. C., 1989; pp. 229-141.
5.	Ames, J. M. *Trends in Food Science and Technology* **1990,** *1,* 150-154.
6.	Reineccius, G. A. In *The Maillard Reaction in Food Processing, Human Nutrition and Physiology;* Finot, P. A.; Aechbacher, H. U.; Hurrell, R. F. and Liardon, R., Eds.; Birkhäuser Verlag: Basel, 1990; pp. 157-170.
7.	Lingnert, H. In *The Maillard Reaction in Food Processing, Human Nutrition and Physiology;* Finot, P. A.; Aechbacher, H. U.; Hurrell, R. F. and Liardon, R., Eds.; Birkhäuser Verlag: Basel, 1990; pp. 171-185.
8.	Ames, J. M. and Nursten, H. E. In *Trends in Food Science;* Lien, W.S. and Foo, C.W. Eds.; Singapore Institute of Food Science and Technology, 1989; pp. 8-14.
9.	Ames, J. M., Apriyantono, A. and Arnoldi, A. *Food Chem.* **1993,** *46,* 121-127.
10.	Bailey, R. G., Ames, J. M. and Monti, S. M. *J. Sci. Food Agric.,* **1995a,** submitted.
11.	Banks, S. B., Ames, J. M. and Nursten, H. E. *Chem. Ind.,* **1988,** 433-434.

12.   Bailey, R. G., Nursten, H. E. and McDowell, I. *J. Chromatogr.*, **1992**, *542*, 115-128.
13.   Snyder, L. R., Stadalius, M. A. and Quarry, M. A. *Anal. Chem.*, **1983**, *55*, 1412A-1430A.
14.   Bailey, R. G., Ames, J. M. and Monti, S. M. *Hewlett-Packard Applications Note*, **1995b**, submitted.
15.   Westphall, G., Kroh, L. and Föllmer, U. *Nahrung*, **1988**, *32*, 117-120.
16.   Hayase, F. and Kato, H. *Agric. Biol. Chem.*, **1985**, *49*, 467-473.
17.   Ames, J. M. and Apriyantono, A. *Food Chem.*, **1993**, *48*, 271-277.

Chapter 2

# Advanced Maillard Products of Disaccharides: Analysis and Relation to Reaction Conditions

Monika Pischetsrieder and Theodor Severin

Institut für Pharmazie und Lebensmittelchemie der Universität München, Sophienstrasse 10, 80333 München, Germany

The degradation of 1,4-glycosidic linked disaccharides, such as maltose and lactose under the condition of the Maillard reaction has been investigated. Two aminoreductones could be isolated which are derived from the 4-deoxyglucosone. Breakdown via the 1-deoxyglucosone pathway leads to 11 typical disaccharide-products. These compounds are not found in glucose reaction mixtures. All these substances can be analyzed using an HPLC system. The correlation between product composition and reaction temperature, time and pH-value has been investigated.

During the Maillard reaction, reducing sugars react with amino groups of proteins or amino acids which results in the formation of a wide range of products. These compounds are of significant importance for nutrition and food processing (1). The nutritional value of food can be reduced by the loss of essential amino acids, decrease of digestibility and formation of toxic and carcinogenic compounds. On the other hand, the so called nonenzymatic browning of food also leads to flavoring, colored and antioxidative products which improve the quality of food and reduce oxidative spoilage.

Several efforts have been made to estimate these quality features or determine heat treatment of food by analyzing Maillard products. Especially Amadori compounds or their analogues furosine and pyridosine (2,3), as well as advanced products, such as pyrraline, have been suggested for this use (4).

In this paper we concentrated on Maillard products derived from 1,4-glycosidic linked disaccharides, such as maltose or lactose. Carbohydrates with a 1,4-glycosidic link are of considerable relevance for food chemistry. For example, starch is one of the main components of many food stuffs. Maltose is an important degradation product of starch and occurs as such particularly in malted food, whereas the milk sugar lactose is found in dairy products.

0097–6156/96/0631–0014$15.00/0

**Degradation Pathway of Disaccharides**

When reducing sugars react with primary or secondary amines, the initial products are N-substituted-1-amino-1-deoxyketoses (Figure 1). These so called Amadori products can be isolated and detected from a large variety of different sugars and amino acids (*1*). However, they are not stable during prolonged storage or heating and undergo degradation to give several different dicarbonyl compounds. The most common are 1-deoxyglucosone (1-DG), 3-deoxyglucosone (3-DG) and 1-amino-1,4-dideoxyglucosone (4-DG). The ratio of the deoxyglucosones are particularly dependent on pH-value. The degradation of these intermediates follows three different pathways.

**3-Deoxyglucosone pathway.** The formation of 3-DG is favored under slightly acidic conditions, whereas it's of less importance at neutral or alkaline pH-range (*5*). Further degradation of 3-DG leads to two known advanced products: HMF and pyrraline (6) (Figure 2). HMF is mainly formed after treatment of reaction mixtures with strong acids and it is of minor interest as a naturally occurring Maillard product, whereas pyrraline could be detected in heated food, where it can be formed from mono- and disaccharides (*3*). This means that neither the extent of formation nor the further composition of 3-DG is influenced by the presence of a glycosidic linked sugar in the molecule, but it can be equally formed from glucose or maltose/lactose.

**4-Deoxyglucosone pathway.** 4-DG appears to be produced at slightly higher pH-values and as a degradation product of 1,4-glycosidic linked disaccharides, such as maltose or lactose. Thus far, only little is known about advanced products which are derived from 4-DG. Recently we were able to isolate and identify two aminoreductones which are obviously formed from this intermediate (*6*): The 4-deoxyaminoreductone (4-DA) is a tautomer form of 4-DG (Figure 3). 4-Deoxyosone has not been isolated in its free form, so that 4-DA must be considered as its more stable tautomer.

The other new compound, 5,6-dihydro-3-hydroxypyridone (DHHP), is an isomer with pyridone structure. The formation of both products are favored at relatively low temperatures (e.g. room temperature) and short reaction times (30 min), whereas during prolonged heating they degrade to unidentified products. 4-DA and DHHP were isolated from several maltose mixtures, but could not be detected when the reaction starts from glucose. It can be assumed that 4-DA and DHHP are derived only from 1,4-glycosidic linked disaccharides, since the formation of the intermediate 4-DG is favored from this type of sugar (*7*). Both products can be easily oxidized and may contribute to the antioxidative effect of Maillard compounds in food.

**1-Deoxyglucosone pathway.** Degradation of Amadori products under neutral conditions leads to the formation of 1-DG. The next steps of the breakdown are cyclization and enolization of 1-DG, resulting in a 5- or 6-membered ring. Starting from these intermediates, the degradation of maltose and lactose differ from that of glucose.

For the case of glucose, the pyranoid product eliminates water to give a γ-pyranone, which is a typical advanced product of monosaccharides. On the other

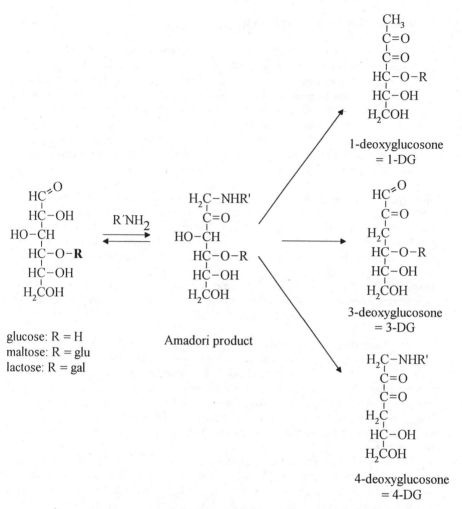

Figure 1. Initial steps of the Maillard reaction of mono- and disaccharides

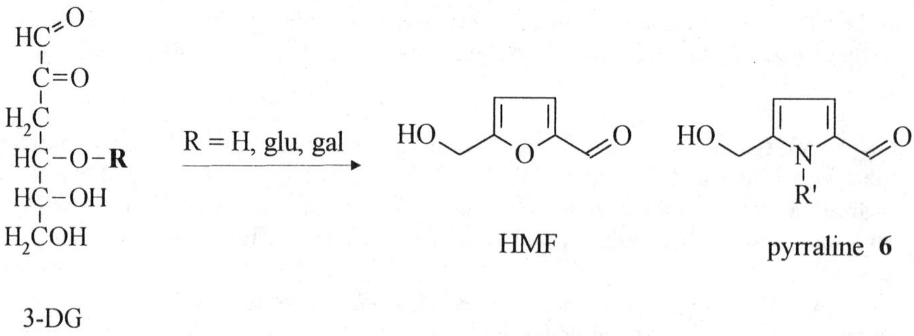

Figure 2. 3-Deoxyglucosone pathway

hand, this reaction is not facilitated if a sugar is bound to position 4, so that a β-pyranone (1) is formed from maltose or lactose (*8*) (Figure 4).

A similar rationale can be made for the 5-membered intermediate. The glucose derivative dehydrates to a 4-furanone. 4-Furanone is a relatively labile substance which can be found only in monosaccharide reaction mixtures. In the furanoid structure which is derived from disaccharides, the bound sugar prevents this reaction and the molecule stabilizes itself by forming a 3-furanone (Figure 4). Recently we were able to isolate 3-furanone (7) and establish its structure by spectroscopic data (*9*).

So it can be concluded that in the case of 1-DG, the glycosidic linked sugar in maltose or lactose causes a degradation pathway which is significantly different from glucose and which gives rise to typical disaccharide products.

**Further Degradation of Disaccharides**

β-Pyranone is not a stable end product of the maltose degradation, but reacts in the presence or absence of amines to give a wide range of advanced products (*8*). In aqueous solution β-pyranone isomerizes to a certain extent to cyclopentenone (2) (Figure 5). The reversible transformation represents an intramolecular aldol type reaction.

During prolonged heating β-pyranone and cyclopentenone eliminate water and are irreversibly converted into the more stable glycosylisomaltol (3). 1, 2 and 3 are stable enough to be isolated or detected, however in the presence of amines they easily condense to nitrogen containing products. β-Pyranone and cyclopentenone react mainly with amines to give acetylpyrrol (5) (Figure 6). Similar to isomaltol, acetylpyrrol can be considered as a late and fairly stable Maillard product of maltose or lactose. Furthermore, isomaltol is able to add an amine, resulting in a mixture of pyridinium betaine (4) and furanoneamine (8). Pyridinium betaine may also be formed by the incorporation of amine into an earlier intermediate.

Further decomposition results in the stabilization of the fairly labile pyridinium betaine (Figure 7). Elimination of the sugar residue leads to the formation of a pyridone derivative, a stable compound which can be considered as one of the end products of the Maillard reaction of maltose. Under slightly alkaline conditions an intramolecular rearrangement of the sugar residue can also take place forming a pyridone glycoside. Furthermore, the sugar can be substituted by a second molecule of amine to give pyridone imine. Assuming that the reaction involves two lysine side chains, pyridone imine represents a possible crosslink product of proteins.

**Analysis of Reaction Products**

In preliminary experiments, model reaction mixtures were investigated. Maltose was reacted with amines such as propylamine or α-N-acetyllysine. To detect all the previously mentioned maltose products at the same time, an HPLC system can be used with a water-acetonitrile gradient (*8*). To avoid chromatographic problems with metal complexing substances like pyridone or pyridone imine, common RP-18 columns are to be replaced by Ultracarb material. Applying diode-array-detection, peaks can be identified by their characteristic UV-spectra which allows an unequivocal analysis of the substances.

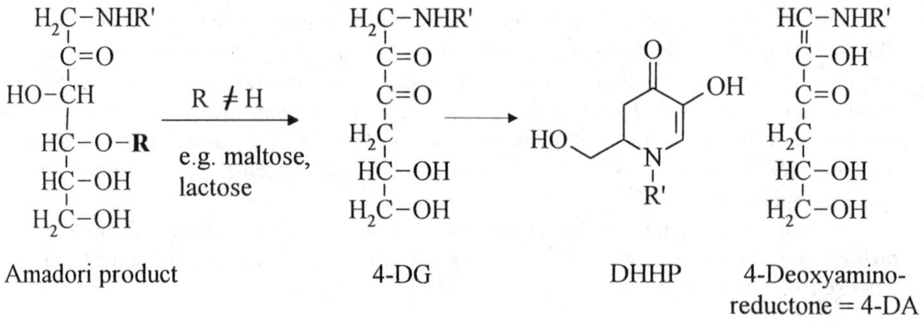

Figure 3. 4-Deoxyglucosone pathway

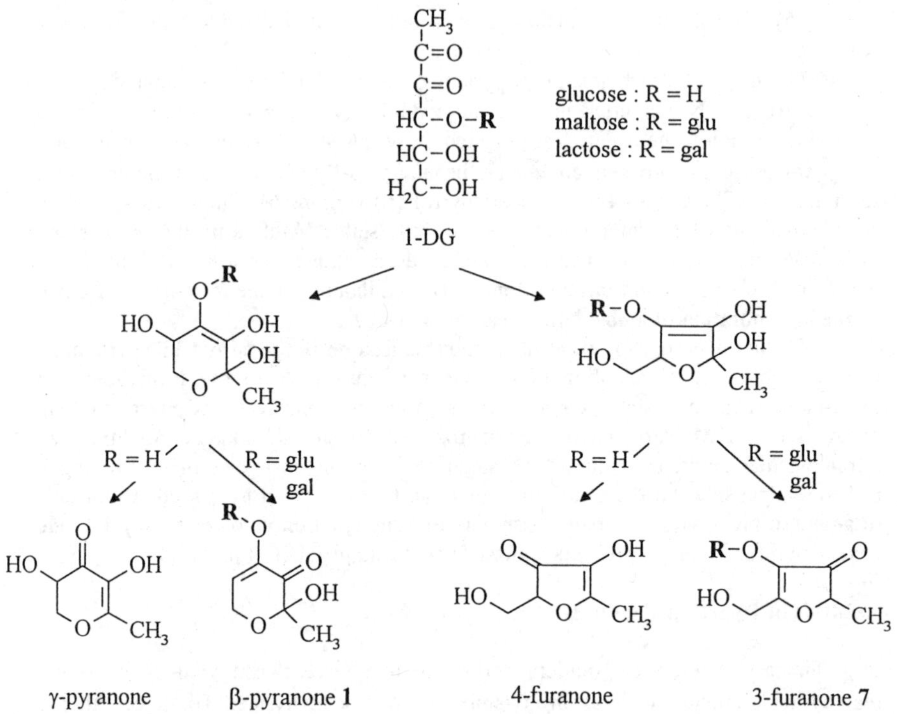

Figure 4. Glucose- and maltose products of the 1-Deoxypathway

β-pyranone **1**                      cyclopentenone  **2**

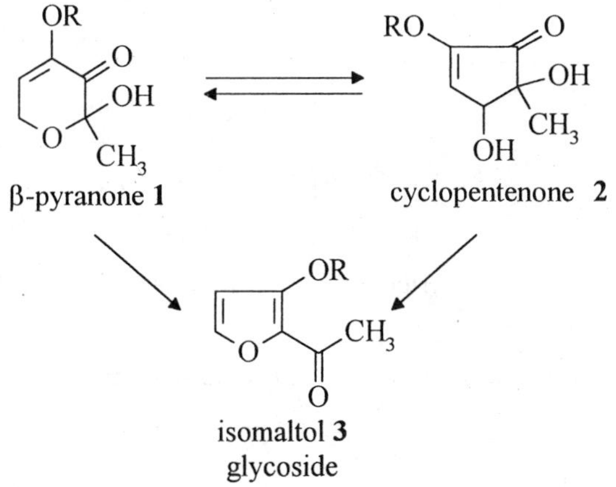

isomaltol **3**
glycoside

Figure 5. Nitrogenfree products deriving from β-pyranone

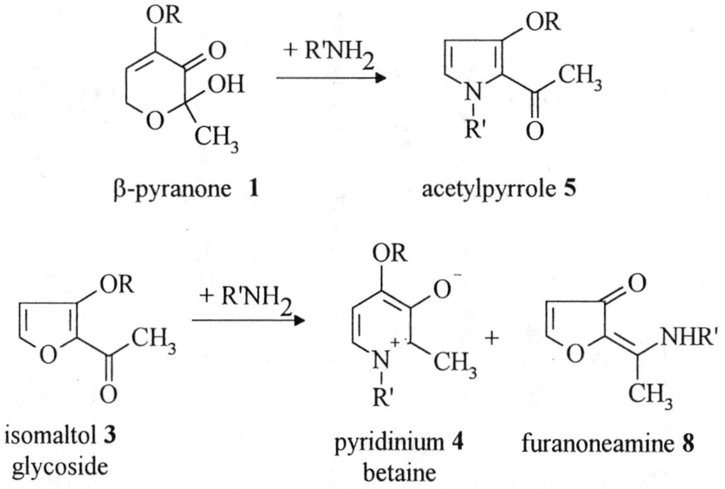

β-pyranone  **1**                      acetylpyrrole **5**

isomaltol **3**              pyridinium **4**      furanoneamine **8**
glycoside                       betaine

Figure 6. Reaction of β-pyranone and glycosyl isomaltol with amines

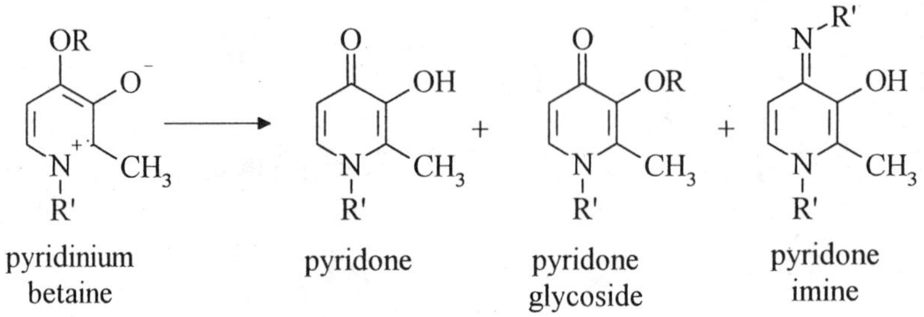

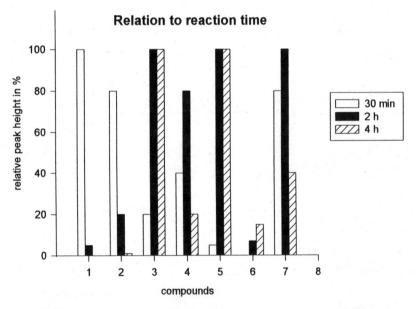

Figure 7. Degradation of pyridinium betaine

Figure 8. Correlation between product composition and reaction time

A number of experiments has shown so far that lysine side chains of proteins react in a similar way as propylamine or free lysine. So it can be assumed that results which were obtained from model mixtures can be transferred to reactions of proteins or even to the conditions in food.

**Correlation between Product Composition and Reaction Time**

After short time heating of maltose in the presence of amines (30 min. at 100°C) the early maltose products predominate (Figure 8). 1-DG decomposes mainly to the 3-furanone, β-pyranone and its isomerization product cyclopentenone. Further degradation products such as glucosyl isomaltol or nitrogen containing compounds like pyridinium betaine and acetylpyrrol are detected in smaller amounts. The latter request incorporation of amines into early intermediates like **1**, **2** or **3**.

Prolonged heating (e.g. 2 h at 100°C) decreases the amount of early products due to the more stable compounds **3**, **4** and **5**. This tendency becomes even more obvious when the reaction is continued for longer (4h).

**Correlation between Product Composition and pH**

Considering the relevance for food processing, a relatively narrow pH range was investigated (Figure 9): pH 7 for neutral, pH 5 for slightly acidic and pH 9 for slightly alkaline conditions.

At lower pH value the 3-deoxypathway is favored over the 1-deoxypathway. Pyrraline is by far the main product under these conditions, whereas of the 1-DG-derived compounds only glucosyl isomaltol and acetylpyrrol can be detected in significant amounts. In a neutral mixture pyrraline is found as a minor compound and at pH 9 it couldn't be detected.

These experiments confirm a correlation which was previously described using other methods: the 1-deoxypathway is the most important mechanism regarding neutral pH range. Decreasing the pH value favors the 3-deoxypathway. The 4-deoxypathway becomes significant for disaccharides under more alkaline conditions (data not shown).

Furthermore, at pH 9 the equilibrium between β-pyranone and cyclo-pentenone is significantly shifted in favor to the latter. This fact can be explained, because the isomerization proceeds via an intramolecular aldol reaction which is generally favored at higher pH values (Figure 10). Moreover furanoneamine (**8**), a condensation product of amine and glucosyl isomaltol, can be detected in considerable amounts under these conditions.

**Correlation between Product Composition and Reaction Temperature**

Low temperatures (70°C) favor the formation of early products (**1** and **7**) and slow down their degradation to more stable compounds (Figure 11). On the other hand higher temperatures lead to more rapid production of early Maillard reaction products. At 130°C for example, β-pyranone, cyclopentenone and 3-furanone are barely detectable, whereas late products such as glucosyl isomaltol, acetylpyrrol and pyrraline become predominant.

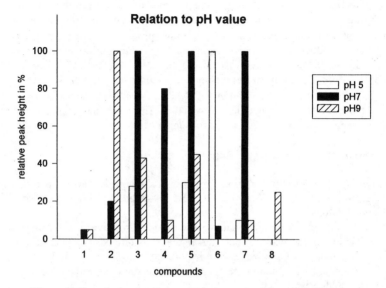

Figure 9. Correlation between product composition and pH-value

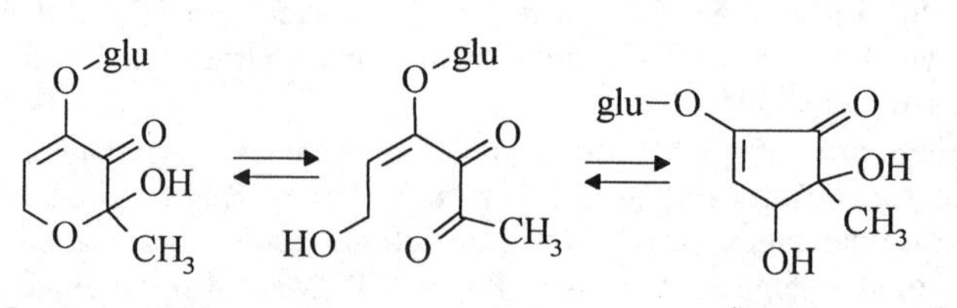

β-pyranone **1**                                    cyclopentenone **2**

Figure 10. Reversible transformation between β-pyranone and cyclopentenone

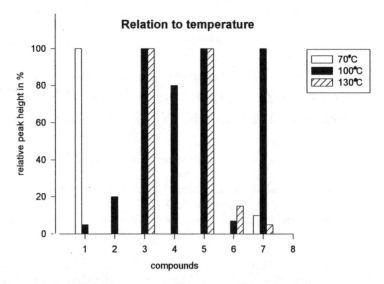

Figure 11. Correlation between product composition and reaction temperature

## Discussion

Reaction mixtures of maltose or lactose with monomer amines were investigated. The products could be separated by HPLC and correlation with reaction conditions was determined. Some of the advanced Maillard products of disaccharides can be detected in heated food like pasta or bread (*10*). Nitrogen-free compounds can be analyzed after appropriate extraction. However, most of the nitrogen-containing advanced products in food are the result of incorporation of lysine side chains and are consequently protein bound. Analysis must therefore be preceded by a mild protein hydrolysis, e.g. by enzymatic means (*3*). Therefore subsequent experiments are in progress analyzing advanced disaccharide products in food and investigating the correlation of their presence and yield with heating conditions.

## Literature Cited

1. Ledl, F.; Schleicher. E. *Angew. Chem.* **1990,** *102*, 597-626.
2. Hartkopf, J.; Ebersdobler, H. F. *J. Chromat.* **1993,** *635*, 151-154.
3. Henle, T.; Klostermeyer, H. *Z. Lebensm. Unters. Forsch.* **1993,** *196*, 1-4.
4. Nakayama, T.; Hayase, F.; Kato, H. *Agric. Biol. Chem.* **1980,** *44*, 1201-1202.
5. Beck, J.; Ledl, F.; Severin, T. *Carbohydr. Res.* **1988,** *177*, 240-243.
6. Schoetter, C.; Pischetsrieder, M.; Severin, T. *Tetrahed. Lett.* **1994,** *40*, 7369-7370.
7. Beck, J.; Ledl, F.; Severin, T. *Z. Lebensm. Unters. Forsch.* **1989,** *188*, 118-121.
8. Kramhoeller, B.; Pischetsrieder, M.; Severin, T. *J. Agric. Food Chem.* **1993,** *41*, 347-351.
9. Pischetsrieder, M.; Severin, T. *J. Agric. Food Chem.* **1994,** *42*, 890-892.
10. Resimi, P.; Pellegrino, L. In *Maillard Reaction in Chemistry, Food, and Health,* Labuza, T.; Reineccius, G.; Monnier, V.; O'Brien, J.; Baynes, J., Eds.; The Royal Society of Chemistry: Cambridge, UK, **1994,** p 418.

# Chapter 3

# The Use of Aminoguanidine To Trap and Measure Dicarbonyl Intermediates Produced During the Maillard Reaction

**M. S. Feather[1], V. Mossine[1], and J. Hirsch[2]**

[1]Department of Biochemistry, University of Missouri, Columbia, MO 65211
[2]Institute of Chemistry, Slovak Academy of Sciences, Dubravska cesta 9, 842 38 Bratislava, Slovakia

Aminoguanidine, an inhibitor of the Maillard reaction functions by rapidly reacting with dicarbonyl intermediates, which are produced from Amadori compounds, converting them into stable 5- and 6- substituted 3-aminotriazine derivatives. The latter can be easily identified and quantitated (by GLC), thus allowing an assessment to be made, not only of the rate of degradation of Amadori compounds to dicarbonyl intermediates, but the type of intermediate that is formed in the reaction as well. Using a number of Amadori compounds prepared from glucose and amino acids, the results show that two different dicarbonyl compounds are produced in the reaction, namely the expected 3-deoxy-1,2-dicarbonyl derivative, but also the 1-deoxy-2,3-dicarbonyl intermediate. The latter represents an intermediate in the formation of a number of methyl furanones which are food flavor and aroma constituents.

The Maillard reaction represents a complex of degradative reactions that involve the initial interaction of a carbonyl-containing compound (usually a reducing sugar) with an amino group (usually an amino acid or a protein molecule). The resulting degradation reaction gives rise to UV absorbing materials, the eventual disappearance of both reducing sugar and amine, the formation of furan derivatives as well as food flavor and aroma constituents, and, lastly, the formation of brown pigments that contain carbon atoms derived from both the sugar and amine (1 - 4). While the entire reaction is not well understood, the initial reactions are reasonably well known and involve the formation of 1-amino-1-deoxy-2-ketose derivatives (Amadori compounds) as shown in Figure 1.

**Degradation of Amadori Compounds to Dicarbonyl Intermediates.**

Amadori compounds are known to be unstable (5) and readily undergo degradation.

0097–6156/96/0631–0024$15.00/0
© 1996 American Chemical Society

Thus, the formation of an Amadori compound represents a pathway by which glucose (an abundant and relatively stable sugar) can undergo degradation at relatively mild conditions. In solution, Amadori compounds are conformationally unstable, as evidenced by NMR spectra, and exist in a number of ring forms, with the β-pyranose form being the most abundant. More recent X-ray studies show that the crystalline modification also is in this tautomeric form (*6*). The exception for this is "difructose glycine", an Amadori compound that contains two 1-deoxy fructose units attached to the amino group of glycine (*7*). This compound was originally used by Anet (*8*) in his pioneering studies of dicarbonyl intermediates produced during the degradation of Amadori compounds and was the compound from which he isolated "3-deoxyglucosone". As might be expected, difructose glycine is unstable and highly reactive. Single crystals of this compound were recently grown by us and the crystal structure obtained using X-ray diffraction techniques. Surprisingly, one of the fructose residues is in the expected β-pyranose form, but the other is in the open chain form. To our knowledge, this represents the only carbohydrate that is known to exist in an acyclic form. This may, in part, explain the high reactivity of this compound.

### Mechanistic Aspect of the Degradation Reaction.

Amadori compounds undergo initial degradation via the formation of deoxy-dicarbonyl intermediates, as shown in Figure 2. 3-Deoxyglucosone was isolated and identified by Anet (*8*) and Kato (*9*) in the early 1960's and represents the most well known of all the putative intermediates. Although unstable, the compound can be synthesized and isolated in relatively large quantities. It has been shown to take part in the crosslinking of proteins at much faster rates than glucose itself (*10*, *11*), is converted into 5-hydroxymethyl-2-furaldehyde in acid solution, and participates in Strecker degradation reactions (*12*). The 1-deoxy-dicarbonyl intermediate (*13*) was identified as a degradation product from an Amadori compound only recently, as was the 1,4-dideoxy derivative (*14*). While these three intermediates are now known to be formed directly from Amadori compounds and, thus, are confirmed as true intermediates in the Maillard reaction, little is known relative to which of these are formed during a "typical" Maillard reaction or their yields. Since they represent intermediates, probably all are formed and react at conditions under which they are produced. Therefore, although only small amounts may be present at a given time, large amounts of the Amadori compound may well degrade via any of them. The 3-deoxyosone serves as a source of hydroxymethyl 2-furaldehyde in acid solution, contributing to some of the UV absorbing compounds produced in the overall reaction. The 1,4-dideoxy intermediate (which remains attached to the amine) undoubtedly serves as the precursor of furosine (*15*), a product of acid treatment of Amadori compounds. Lastly, the 1-deoxyosone (containing both 5 and 6 carbons atoms) probably serves as precursors of some of the methyl furanones produced in heated foods and which represent food flavor and aroma constituents (*16*, *17*). It is generally assumed that the above intermediates are key in the overall reaction, in that they participate in Strecker degradation reactions, serve as precursors of furan and furanone derivatives, and further react with proteins to cause extensive modifications such as crosslinking and the development of fluorescent bodies. Clearly other factors participate in as yet to be understood fashions. Free radicals,

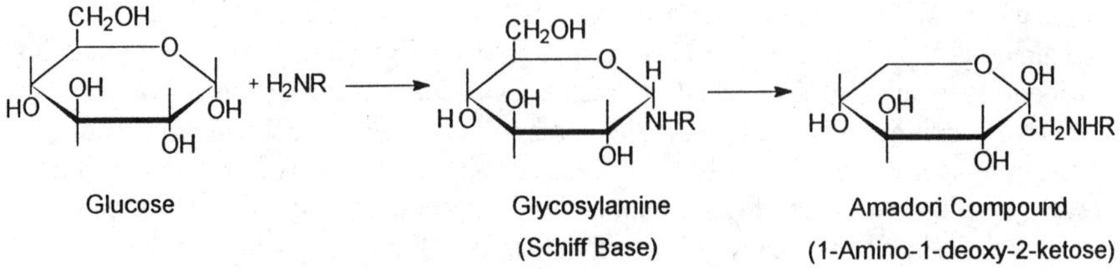

Figure 1. Formation of Amadori compounds from glucose and amine

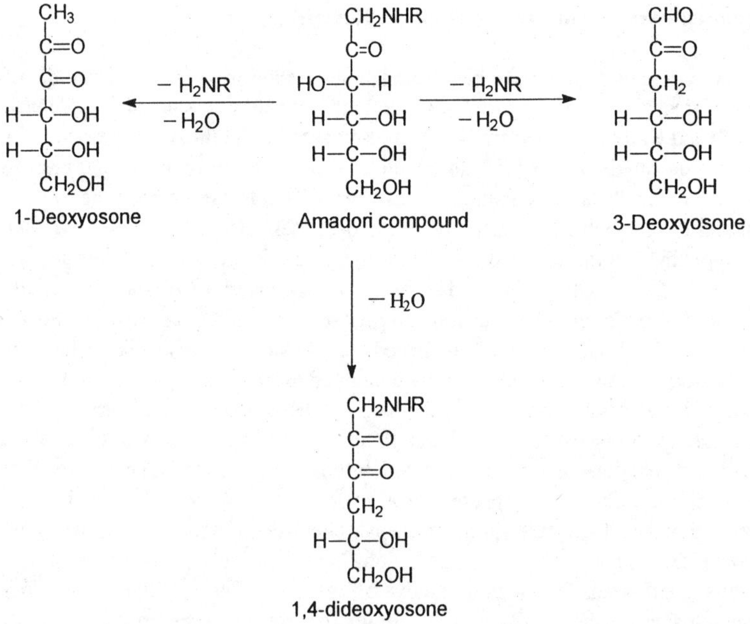

Figure 2. Further reactions of Amadori compounds

clearly are important in the reaction as well, as evidenced by the acceleration of Maillard reactions in the presence of transition metal ions and the presence of oxygen, which contribute to the formation of such radicals. In addition, probably fragmentation of sugar molecules, or Amadori compounds by reverse aldol cleavage reactions may well also constitute a factor in this reaction, since such reactions give rise to carbonyl and dicarbonyl compounds as well. Oxidative cleavage reactions also probably play a role in the formation of some end products. For example, the isolation of ε-N-carboxymethyl lysine in biological materials (*18*) as well as from an Amadori compound, suggests that it is produced via an "oxidative" cleavage between C-2 and C-3 of the Amadori compound. This almost surely represents an oxidative reaction involving free radicals. This discussion, however, will be limited to dicarbonyl compounds that are derived from Amadori compounds.

**Inhibition of the Maillard Reaction.**

The food industry has had a longstanding interest in inhibiting the Maillard reaction, in order to prevent color formation, as well as the production of off flavors. Inhibition has proved to be a difficult task. At best such reactions can be minimized by pH and temperature control, or by the addition of bisulfite, a substance that, presumably, reacts with carbonyl groups and prevents them from participating in Maillard reactions. Recently, aminoguanidine (*19*) has been shown to be a effective inhibitor for *in vivo* Maillard reactions. There is considerable debate as to the mechanism for this inhibition. Aminoguanidine appears to inhibit color formation, as well as protein modifications such as crosslinking and the formation of fluorescent materials associated with aging. Aminoguanidine appears to be relatively nontoxic and is now being tested as a possible drug for use in the treatment of diabetic complications. It is known that aminoguanidine reacts with carbonyl compounds to give hydrazones (*20*), and, as a result, we hypothesized that it may well react with the dicarbonyl intermediates formed initially in Maillard reactions. This would allow the "trapping" of such intermediates, thus stabilizing them and preventing them from participating in further reactions. In some initial experiments, we examined the reaction of aminoguanidine with 3-deoxy-D-*threo*-pentos-2-ulose (3-deoxyxylosone), and found that it reacted almost instantaneously and irreversibly to give the stable 5- and 6-substituted 3-aminotriazine derivatives (*21*), as shown in figure 3. In further experiments, we examined this reaction using several other dicarbonyl compounds synthesized in our laboratories (3-deoxyglucosone, glucosone and xylosone) and found that the reaction is analogous, i.e., in all cases, the reagent reacts rapidly and irreversibly (*22*) to give 3-aminotriazine derivatives (figure 3). The only difference in the reactions were that, the 3-deoxy derivatives give both the 5- and 6- substituted isomers, while the osones themselves give only the 5 substituted isomer. We have spent considerable time and effort in characterizing the products of the reaction and have unequivocally characterized the triazines by NMR, mass spectrometry, and, in one case (3-deoxyglucosone), the structures were confirmed by X-ray diffraction studies (*23*). Since the triazines accumulate in the reaction and are formed rapidly, they can be used to measure the rate of formation of dicarbonyl intermediate that is formed in a Maillard reaction. The triazines are conveniently measured by converting them to the trimethylsilyl (TMS)

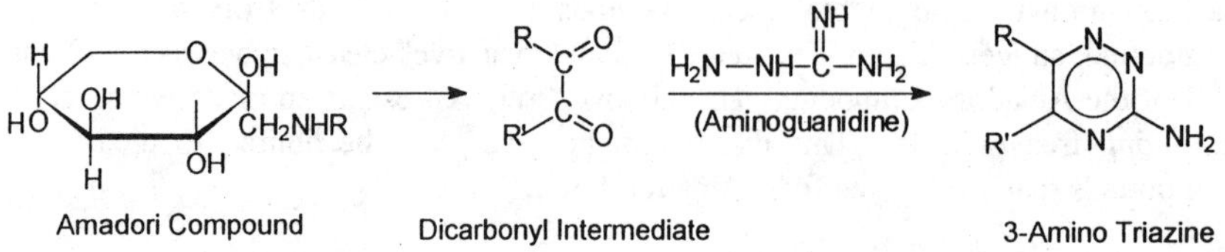

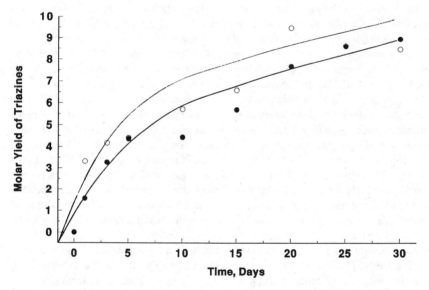

Amadori Compound        Dicarbonyl Intermediate        3-Amino Triazine

Figure 3. Formation of triazine derivatives from dicarbonyl intermediates

Figure 4. Molar yields of triazines produced from fructoseglycine at the pH corresponding to its pKa (pH = 8.2). Open circles = 1-deoxyglucosone, closed circles = 3-deoxyglucosone.

derivative and then using capillary GLC techniques as the method of detection. A problem is that only the triazines derived from the 3-deoxyosone intermediates were available, and, in order to measure any others that may be formed in Maillard reactions, it is necessary to prepare standard materials. The most interesting intermediate would be the 1-deoxyosone intermediate, but, although it has been reported as a component of Maillard reactions, it has never been isolated in pure and underivatized form. A synthesis of the 5,6-O-isopropylidine derivative was reported by This Laboratory some years ago (*24*), and this provides a route to the standard triazine derivative that was needed for the analyses (*25*). The strategy for this study involved incubation of Amadori compounds at specified conditions in the presence of aminoguanidine. The assumption was that either dicarbonyl intermediate, when formed from the Amadori compound, would immediately react with aminoguanidine to give a stable triazine, which could be quantitated as a function of time by conversion to the TMS derivative and measurement by GLC. Figure 3 shows the assumed scheme for the overall reactions. The data in Figures 4 - 6 show plots of the yields of triazines and, hence, dicarbonyl compounds produced from several Amadori compounds as a function of time. Of interest is the fact that, for the two monosubstituted Amadori compounds (fructose lysine and -glycine, respectively), the yields of 1-deoxyosone and 3-deoxyosone are about equivalent and, together, comprise a nearly 10 - 20% total yield after 30 hours of reaction. The presence of 1-deoxyosone was surprising, in that it is normally assumed to be a minor contributor for Maillard reactions. Also surprising is

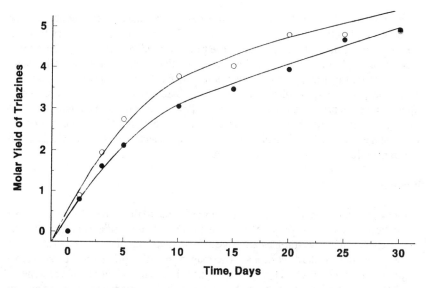

Figure 5.   Molar yields of triazines produced from fructoselysine at the pH corresponding to its pKa (pH = 5.2). Open circles = 1-deoxyglucosone, closed circles = 3-deoxyglucosone.

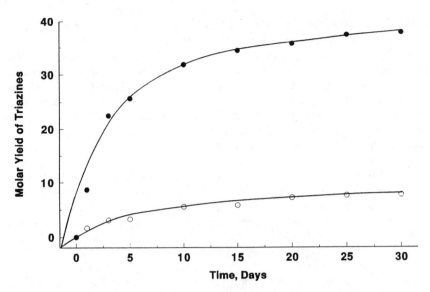

Figure 6. Molar yields of triazines produced from difructoseglycine at the pH corresponding to its pKa (pH = 9.0)  Open circles = 1-deoxyglucosone, closed circles = 3-deoxyglucosone.

the fact that, for difructose lysine, the yields of 3-deoxyosone are considerably higher, reaching nearly 40% at 30 hours of reaction.  This may reflect that fact that this compound exists in solution with one of the fructose residues in the open chain form, a form that is generally assumed to be a more reactive form of the sugar.  The yields also reflect that fact that, at these conditions, the fructose residues of this Amadori compound undergo 1,2-enolization to a greater extent than 2,3-enolization.  These experiments were performed at a pH corresponding to the pKa of the amino group to which the sugars are attached.  It remains to be seen if solution pH has an effect on the course of the reaction, i.e., if a higher or lower pH can change the relative yields of the dicarbonyl intermediates formed in the reactions.

**Summary.**

In summary, aminoguanidine is a potent inhibitor of the Maillard reaction and appears to function by reacting with highly reactive dicarbonyl intermediates and converting them to stable, unreactive triazine derivatives, thus preventing them from participating further in overall Maillard reactions.  The reagent has also proved to be useful as a "trapping" reagent which allows, for the first time, an evaluation of the type of dicarbonyl intermediate that is produced during the degradation of an Amadori compound and, thus, has contributed significantly to our understanding of this reaction.

**Acknowledgments.**

Much of the experimental work done in this project was supported by the Juvenile Diabetes Foundation (Grant number 192198). The Bruker NMR spectrometer used to collect data which are used herein was purchased, in part, via National Science Foundation grant 8908304. X-ray structures were obtained using a difractometer purchased, in part, by National Science Foundation grant CHE 90 - 11804.

**Literature Cited.**

1. Amino-carbonyl reactions in food and biological systems; Fujamaki, M.; Namiki, M.; Kato H., Eds. Kadansha: Tokyo, **1985**; Vol. 13
2. Ledl, F.; Schleicher, E., *Angew. Chem.* **1990**, 29, 565-706
3. Progress In Food And Nutritional Science--Maillard Reactions In Food; K. Eriksson, Ed Pergamon Press: New York, N.Y., **1981**, Vol 5
4. The Maillard Reaction in Aging, Diabetes, and Nutrition; Baynes, J. W.; Monnier, V. M. Eds.; Progress In Clinical And Biological Research, Alan R. Liss, Inc., New York, N.Y., **1989**; Vol. 304
5. Feather, M.S.; Russell, K.R. *J. Org. Chem.* **1969**, 34, 2650-2652
6. Mossine, V. V.; Barnes, C. L.; Glinsky, G. V.; Feather, M.S. *Carbohydr. Res.* **1994** 262, 257 - 270
7. Mossine, V. V.; Barnes, C. L.; Feather, M.S.; *Carbohydr. Lett.* IN PRESS
8. Anet, E.F.L.J. *J. Am. Chem. Soc.* **1960**, 82, 1502
9. Kato, H. *Agric. Biol. Chem.-Tokyo* **1962**, 26, 187-192
10. Kato, H.; Cho, R.K.; Okitani, A.; Hayase, F. *Agric. Biol. Chem.* **1987**, 51, 683-689
11. Kato, H.; Shin, D.B.; Hayase, F. *Agric. Biol. Chem.* **1987**, 51, 2009-2011
12. Ghiron, A.F.; Quack, B.; Mawhinney, T.P.; Feather, M.S. *J. Agric. Food Chem.* **1988**, 36, 673-677
13. Beck, J.; Ledl, F.; Severin, T. *Carbohydr. Res.* **1988**, 177, 240-243
14. Huber, B.; Ledl, F. *Carbohydr. Res.* **1990**, 204, 215-220
15. Finot, P.A.; Deutsch R.; Bujard J *Prog. Fd. Nutr. Sci.* **1981**, 5, 345-355
16. Hicks, K.B.; Feather, M.S. *J. Agric. Food Chem.* **1975**, 23, 957-960
17. Peer, H.G.; van den Ouwelad, G.A.M.; de Groot, C.N. *Rec. Trav. Chim* **1968**, 87: 1011-1016
18. Ahmed, M.U.; Thorpe, R.S.; Baynes, J.W. *J. Biol. Chem.* **1986**, 261, 4889-4894
19. Brownlee, M.; Vlassara, H.; Kooney, A.; Ulrich, P.; Cerami, A. *Science* **1986**, 232 1629 - 1632
20. Wolfrom, M. L.; El Khadem, H.; Alfes, H. *J. Org. Chem.* **1964** 29, 3074 - 3076
21. Hirsch, J.; Baynes, J.W.; Blackledge, J.A.; Feather, M.S. *Carbohydr. Res.* **1991**, 220, C5-C7
22. Hirsch, J.; Petrakova, E.; Feather, M.S. *Carbohydr. Res.* **1992**, 232, 125-130
23. Hirsch, J.; Barnes, C.L.; Feather, M.S. *J. Carbohydr. Chem.* **1992**, 11, 891-901
24. Feather, M.S.; Eitleman, S.J. (1988) *J. Carbohydr. Chem.* **1988**, 7, 251-262
25. Hirsch, J.; Mossine, V. V.; Feather, M. *Carbohydr. Res.***1995**, 273, 171-177

# Chapter 4

# Early Detection of Changes During Heat Processing and Storage of Tomato Products

K. Eichner, I. Schräder, and M. Lange

Institut für Lebensmittelchemie der Universität Münster, Piusallee 7, D–48147 Münster, Germany

During heat processing of tomatoes dependent on the reaction conditions different chemical reactions take place. Pyrrolidone-carboxylic acid formed by cyclization of glutamine arises already during the break process. In the course of concentration and pasteurization processes several Amadori compounds occur. They represent chemical markers for the onset of the Maillard reaction. Very high concentrations of Amadori compounds could be found in dried tomato products, especially fructose-glutamic acid and fructose-pyrrolidone-carboxylic acid. They are also formed during storage at elevated temperatures. Amadori compounds were analyzed by gas chromatography after proper derivatization. During tomato drying also several heterocyclic compounds and Strecker aldehydes having a low sensory threshold are formed; they can be used as chemical markers for the onset of undesirable changes in flavor. These compounds can be analyzed by headspace gas chromatography and simultaneous distillation-extraction.

Tomato processing involves several heat processing steps for obtaining a shelf-stable product having the desired physical and sensory properties. For inactivation of enzymes the tomatoes are crushed at 60 - 70°C (cold break (CB) process) or at 90 - 95°C (hot break (HB) process) (1). During the CB process pectolytic enzymes are inactivated slowly, therefore pectin is hydrolyzed during the heating process to some extent and galacturonic acid (GalA) is set free (1); GalA may participate in the Maillard reaction being a very reactive reducing sugar (2). In the HB process the enzymes are inactivated quickly and pectin is preserved resulting in a higher viscosity and textural stability of the product. After removing seeds and skins the remaining tomato juice is pasteurized or concentrated under reduced pressure at about 80°C yielding tomato paste with different dry matter (DM) contents (double concentrate: 28-30 % DM; triple concentrate: 36 - 40% DM) (*3*).

Tomato powder or tomato flakes are produced by spray-drying or drum-drying of tomato paste (dry matter content: 30-40 %) (*4*); starch may be added before drying in order to facilitate the drying process and to increase product stability.

In order to evaluate the intensity of thermal treatment during heat processing of tomatoes, it seems desirable to characterize the extent of thermal impact during different steps of heat processing by using chemical markers formed by heat induced chemical reactions of reducing sugars and amino acids in the product. Of primary interest in this connection are early Maillard products (Amadori compounds, cf. (*5*) ) and pyrrolidone-carboxylic acid (PCA) formed by cyclization of glutamine under acidic conditions.

## Chemical Markers for Early Detection of Chemical Changes during Tomato Processing

**Methodology.** Reducing sugars, organic acids and Amadori compounds were determined by gas chromatographic separation on capillary columns. First Amadori compounds were extracted from the tomato products with water and bound to a cation exchange resin (Lewatit S 1080 column); sugars, organic acids and fructose-pyrrolidone-carboxylic acid (Fru-PCA, which is less basic than the other Amadori compounds and not bound to the resin) are eluted from the cation exchange column with water. Thereafter, the Amadori compounds are eluted with 0.5 M trichloroacetic acid (TCA); the TCA is removed by extraction with diethylether. Both eluates are concentrated and freeze-dried.

Prior to gas chromatographic analysis reducing sugars and Amadori compounds were converted into the corresponding oximes (syn- and anti-form) with hydroxyl-ammonium chloride in pyridine (30 min, 70°C) (*6*). The oximes and the organic acids present are treated with N,O-bis-(trimethylsilyl)-acetamide and trimethylchlorsilane (30 min, 70°C) leading to the trimethylsilyl derivatives (*6*). The gas chromatographic separation of the reducing sugars, organic acids and Fru-PCA was carried out on a fused silica capillary column covered with methyl silicone (OV-101). Temperature program: 120-280°C with a heating rate of 6°C/min. Figure 1 shows a standard chromatogram of reducing sugars, saccharose, organic acids, PCA and Fru-PCA. Xylitol and trehalose were used as internal standards. Reducing sugars and the Amadori compound appear as characteristic double peaks, corresponding to the syn- and anti-form of the oximes; the first peak represents the syn-form, the second one the anti-form (*6*).

The gas chromatographic separation of the remaining Amadori compounds was performed using a fused silica capillary column covered with methylpolysiloxane(DB-5). Temperature program: 140-300°C with a heating rate of 6°C/min.

Figure 2 shows a standard chromatogram of Amadori compounds occurring in tomato products (xylitol and trehalose as internal standards). Again, each compound appears as a double peak, corresponding to the syn- and anti-form of the oximes.

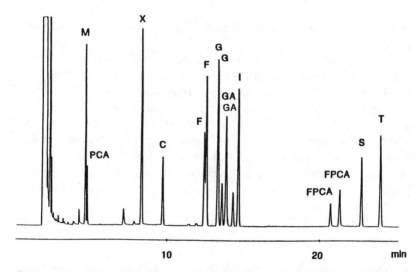

Figure 1. Separation of reducing sugars, saccharose, and organic acids by capillary gas chromatography (standard chromatogram). F = fructose; G = glucose; GA = galacturonic acid; I = inositol; S = saccharose; C = citric acid; M = malic acid; PCA = pyrrolidone-carboxylic acid; FPCA = fructose-pyrrolidone-carboxylic acid. Internal standards: X = xylitol; T = trehalose.

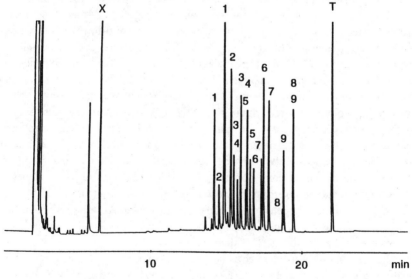

Figure 2. Separation of Amadori compounds by capillary gas chromatography (standard chromatogram) 1 Fru-Ala; 2 Fru-Gly; 3 Fru-Val;4 Fru-Leu; 5 Fru-Gaba; 6 Fru-Ser; 7 Fru-Thr; 8 Fru-Asp; 9 Fru-Glu (Gaba = γ-aminobutyric acid). Internal standards: X = xylitol; T = trehalose.

**Analytical Control of Tomato Juice Processing.**   In Table I characteristic chemical changes during tomato juice processing are shown. The raw material is subjected to a HB process. After a certain holding time at elevated temperatures the product is sieved in order to remove seeds and skins and pasteurized after addition of sodium chloride.

**Table I. Chemical Changes during Processing of Tomato Juice (Concentrations in mmol/100 g dry matter)**

| Processing step: | Fresh product | HB process | Holding period | Pasteurization |
|---|---|---|---|---|
| Dry matter (g/100g) | 6.82 | 7.17 | 6.24 | 6.54 |
| PCA | n.d. | 3.2 | 4.1 | 4.4 |
| Glutamine | 14.4 | 9.5 | 5.0 | 2.0 |
| Fru-PCA | n.d. | n.d. | 0.55 | 0.68 |

(PCA = Pyrrolidone-carboxylic acid; n.d. = non detectable)

As shown in Table I, the first noticeable chemical change occurring during the HB process is the formation of pyrrolidone-carboxylic acid (PCA) by cyclization of glutamine; this reaction is promoted under acidic conditions existing in tomatoes (pH 4.2). At a concentration of 0.05 % in tomato juice PCA creates a tart off-flavor, occasionally described as bitter; however, this concentration limit is not yet reached during tomato juice processing as indicated in Table I. Fructose-pyrrolidone-carboxylic acid (Fru-PCA) being the first noticeable Amadori compound appears not till the holding period and increases during the pasteurization process. Apart from the decrease of glutamine no considerable changes in the concentrations of amino acids could be observed. Therefore it can be concluded that the Maillard reaction plays a minor role under the relatively mild conditions of tomato juice processing.

**Tomato Paste and Tomato Powder Processing Control.**   Evaporation of water in the concentration plant, which operates under reduced pressures, brings about a progressive increase in solids content of the tomato pulp until a paste of the desired density is produced (*3*). Concentration to a final solids content of 28 - 30 % is the most common practice in Europe, and this paste forms the "double concentrate" of commerce. The temperature of the paste must be raised to at least 90°C before it is filled into cans in order to prevent the survival of microorganisms.

Tomato powder usually is produced by spray-drying of tomato paste, whereas tomato flakes are obtained by drum drying of tomato paste, usually after addition of starch.

**Table II: Chemical Composition of Industrially Processed Tomato Paste and Tomato Flakes (g/100 g DM)**

|  | Tomato paste | | Tomato flakes | |
|---|---|---|---|---|
|  | CB | HB | CB | HB |
| Dry matter (DM) | 30.69 | 30.95 | 96.15 | 95.95 |
| Fructose (Fru) | 27.8 | 24.8 | 22.6 | 20.4 |
| Glucose (Glu) | 25.8 | 20.7 | 12.5 | 10.1 |
| Galacturonic acid (GalA) | 0.819 | 0.171 | 0.240 | n.d. |
| Pyrrolidone-carboxylic acid (PCA) | 1.63 | 1.65 | 1.53 | 1.69 |
| Glutamic acid (Glu) | 3.46 | 3.64 | 0.71 | 1.15 |
| γ-Aminobutyric acid (Gaba) | 1.06 | 1.20 | 0.21 | 0.40 |
| Alanine (Ala) | 0.353 | 0.347 | 0.110 | 0.139 |
| Aspartic acid (Asp) | 1.21 | 1.15 | 0.37 | 0.49 |
| Serine (Ser) | 0.205 | 0.192 | 0.079 | 0.102 |
| Threonine (Thr) | 0.172 | 0.163 | 0.119 | 0.081 |
| Fru-Glu | 0.06 | 0.03 | 1.7 | 1.4 |
| Fru-PCA | 0.47 | n.d. | 2.8 | 2.3 |
| Fru-Gaba | 0.11 | 0.19 | 2.4 | 2.3 |
| Fru-Ala | n.d. | n.d. | 0.226 | 0.226 |
| Fru-Asp | n.d. | 0.03 | 2.0 | 1.7 |
| Fru-Ser | n.d. | n.d. | 0.241 | 0.241 |
| Fru-Thr | n.d. | n.d | 0.141 | 0.113 |
| A (420 nm) | 0.061 | 0.071 | 0.196 | 0.161 |

CB: Cold Break; HB: Hot Break; n.d. non detectable
A (420 nm): 0.50 g DM/100 ml solution

Table II shows the chemical composition of tomato paste produced by the cold break (CB) and the hot break (HB) process as well as the composition of tomato flakes produced from CB- and HB-tomato pastes.

It can be seen that the CB-tomato paste contains much more GalA than the HB product, caused by the delayed inactivation of pectolytic enzymes during the CB process. Glutamine is not detectable in both products because of its complete cyclization to PCA during the heating process. On the other side, only a relatively small portion of the amino acids present has been converted to the corresponding Amadori compounds. Figure 3 summarizes the conversion of PCA and the most important amino acids of tomato to Amadori compounds in CB- and HB-tomato paste on a molar basis.

Using Amadori compounds as chemical markers for the onset of the Maillard reaction, it can be concluded that during production of tomato paste the Maillard reaction still has a minor importance.

Compared to these results it was of great interest to see how far the Maillard reaction will proceed under drying conditions taking into account that this reaction has a maximum rate at certain moisture contents which must be crossed during the drying process (7).

As Table II shows, the concentrations of amino acids have been greatly diminished during drum-drying of CB- and HB-tomato paste; at the same time there is a strong increase in the concentration of the corresponding Amadori compounds and of visible browning (A (420 nm) = absorbance value at 420 nm).

Figure 4 again summarizes the conversion of the most abundant amino acids of tomato to Amadori compounds in CB- and HB-tomato flakes on a molar basis.

The Amadori compounds as such do not contribute to sensory changes; however, they generally can be used as chemical markers for an early detection of the Maillard reaction in foods (5, 6, 9). By decomposition of Amadori compounds melanoidins and volatile Strecker aldehydes having a very low sensory threshold value are formed. The ratio between the concentration of Amadori compounds and the intensity of sensory changes depends on the reaction conditions. The CB- and HB-tomato flakes under investigation show only a slight burnt and adstringent off-flavor.

Fructose does not form the corresponding Heyns rearrangement products in detectable amounts; its contribution to browning is comparatively low.

Comparing Figure 4 with Figure 3 it becomes clear that the molar sum of amino acid and Amadori compound remains constant for γ-amino butyric acid (Gaba), whereas in the case of glutamic acid (Glu) this sum is greatly diminished. But it also can be seen in Figure 4 that a great deal of the Glu initially present has been converted to Fru-PCA by cyclization of Fru-Glu, the concentration of PCA remaining about the same in tomato paste and tomato flakes.

**Chemical Changes during Storage of Tomato Powder.**  In order to elucidate storage changes of dried tomato products by chemical markers tomato paste was freeze-dried, adjusted to a water activity ($a_w$) of 0.35 (8) and stored at 40°C. As shown in Figure 5, the concentration of fructose-glutamic acid (Fru-Glu) increases sharply at the beginning of storage and decreases in the course of time after reaching a maximum. On the other hand, the proportion of the relatively stable fructose-pyrrolidone-carboxylic acid (Fru-PCA), the cyclization product of Fru-Glu, increases slowly, whereas the concentrations of the other Amadori compounds listed in Figure 5 after an initial increase remain almost constant. Contrary to the formation of Amadori compounds the onset of visible browning shows an induction period. Figure 5 shows clearly that, for evaluating the course of Maillard reaction due to thermal treatment or storage of tomato powder at elevated temperatures, the sum of the marker substances Fru-Glu and Fru-PCA must be considered (9).

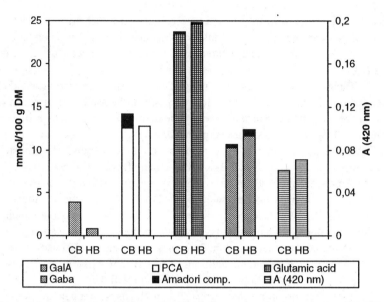

Figure 3. Important constituents of industrially produced tomato paste (CB, HB = cold break resp. hot break process) (cf. Table II for further data).

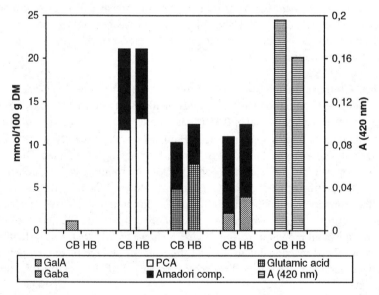

Figure 4. Important constituents of industrially produced tomato flakes (CB, HB = produced from cold break resp. hot break tomato paste) (cf. Table II for further data).

**Chemical Markers for Early Detection of Sensory Changes during Tomato Processing.**

The Maillard reaction which becomes predominant during thermal treatments in food processing (*10*) may create desirable aroma components like in baking and roasting processes (*11-13*), but on the other hand it often causes detrimental sensory changes (*14,15*).

During further progress of the Maillard reaction Amadori compounds being the first stable intermediates of this reaction decompose via 1,2- or 2,3-enolization to yield 3-deoxyosones or 1-deoxyosones, respectively (*5*). 3-Deoxyosones cyclize to form hydroxymethylfurfural (HMF) which can be used as a marker for detecting Maillard reactions in acid foods (*15*). At higher temperatures the "1-deoxyosone-pathway" prevails (*16*). Figure 6 shows the decomposition pathways of the 1-deoxyosone which is formed by 2,3-enolization of Amadori compounds followed by ß-elimination of the amino acid residue (*16*).

By retro-aldol cleavages of the 1-deoxyosone short-chain dicarbonyls like diacetyl, methylglyoxal or hydroxydiacetyl are formed which - like the deoxyosones - may participate in the Strecker degradation of amino acids, thus creating volatile flavor components (*17,18*). On the other hand, the 1-deoxyosone undergoes several cyclization reactions yielding e.g. acetylfuran, maltol, isomaltol, furaneol and 5-hydroxy-5,6-dihydromaltol (DHM) (*16*).

In order to demonstrate the effect of heat impact during drum-drying of tomato paste on the formation of volatile compounds, their concentrations in drum-dried tomato flakes and in tomato paste, from which the flakes were produced, were compared. Table III shows some highly volatile compounds analyzed in tomato paste and tomato flakes by head-space-gas chromatography. The gas chromatographic separation was carried out on a 60 m fused silica capillary, covered with Stabilwax. Temperature program: 40-220°C with a heating rate of 4°C/min, holding time: 20 min. The aldehydes shown in Table III were formed by Strecker degradation of the amino acids alanine, valine, isoleucine and leucine; dimethyl sulfide may be formed by thermal degradation of S-methyl-methionine.

Table III shows that the concentrations of all listed volatile compounds are far above the sensory threshold values (determined in water) thus strongly contributing to flavor; they may contribute to off-flavor, if they exceed certain concentrations. Therefore the Strecker aldehydes and dimethyl sulfide listed in Table III can be used as very sensitive marker substances for sensory changes during heat processing of tomatoes, especially 2- and 3-methylbutanal, since their concentrations are more than hundredfold higher in tomato flakes than in tomato paste.

In Table IV thermally generated Maillard reaction products having a medium volatility are shown; they were isolated from the tomato products by simultaneous distillation-extraction (SDE) using methylene chloride as a solvent and separated by gas chromatography on a 60 m fused silica capillary, covered with Stabilwax. Temperature program: 40°C for 5 min; 40-220°C with a heating rate of 5°C/min, holding time: 30 min.

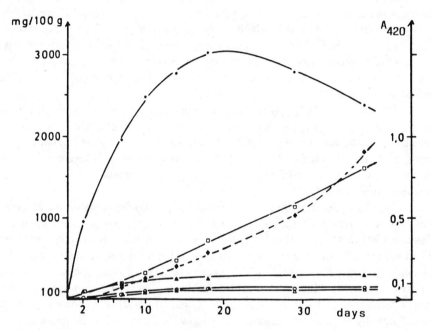

Figure 5. Storage of freeze-dried tomato powder at 40°C and a water activity (a$_w$) of 0.35. ● Fru-Glu, ◆ Fru-PCA, ▲ Fru-Asn, ○ Fru-Thr, Δ Fru-Ser, ❏ Browning (A(420 nm)); the absorbance values refer to an extract of 0.5 g sample material in 25 ml water.

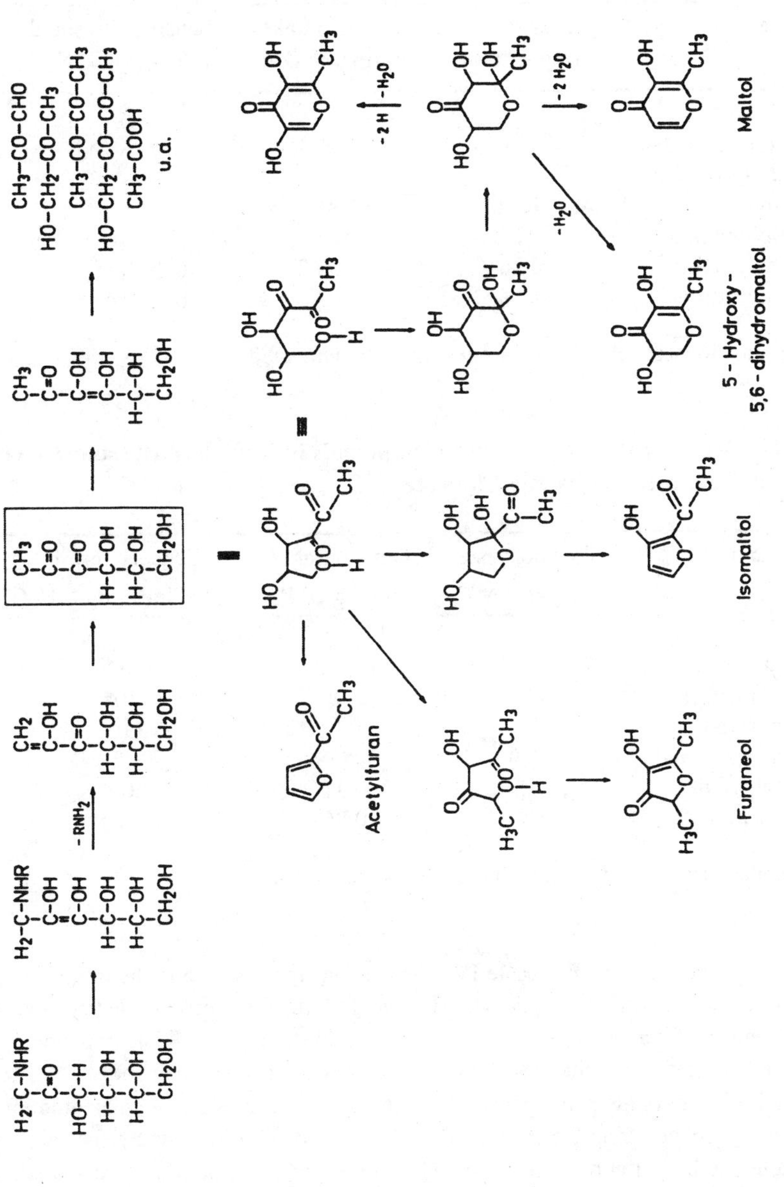

Figure 6. Decomposition pathways of the 1-deoxyosone *(16)*.

**Table III. Determination of Volatile Compounds in Tomato Paste and Tomato Flakes by Static Head-Space-Gas Chromatography**

| Compound | Tomato paste (mg/kg DM) | Tomato flakes* (mg/kg DM) | Sensory threshold values (mg/L $H_2O$) |
|---|---|---|---|
| Acetaldehyde | 6.4 | 75.1 | 0.015 |
| 2-Methylpropanal + Acetone | 18.0 | 81.9 | |
| 2-Methylbutanal | 0.21 | 27.6 | $2 \times 10^{-3}$ |
| 3-Methylbutanal | 0.26 | 31.2 | $0.2 \times 10^{-3}$ |
| Dimethyl sulfide | 10.6 | 63.9 | $0.3 \times 10^{-3}$ |

* containing 56 % starch; all data are related to tomato solids

**Table IV. Determination of Volatile Compounds in Tomato Paste and Tomato Flakes by SDE with Methylene Chloride**

| Compound | Tomato paste (mg/kg DM) | Tomato flakes (mg/kg DM) | Sensory threshold values (mg/L $H_2O$) |
|---|---|---|---|
| Furfural | 3.34 | 51.10 | 3.0 |
| 5-Methylfurfural | 0.06 | 7.95 | 10* |
| 2-Acetylfuran | n.d. | 1.30 | 110* |
| Phenylacetaldehyde | 2.19 | 48.0 | $4 \times 10^{-3}$ |
| 2-Acetylpyrrole | n.d. | 0.31 | 200* |
| 2-Formylpyrrole | n.d. | 0.85 | |

n.d. = non detectable.  * determined in orange juice

All compounds listed in Table IV show a strong increase due to the drum drying process. The concentrations of phenylacetaldehyde formed by Strecker degradation of phenylalanine lie far above the sensory threshold values; therefore this compound is a useful marker for sensory changes due to heat processing of tomatoes, too.

Table V shows thermally generated Maillard reaction products with medium and low volatility isolated from the tomato products by extraction with methylene chloride and cleaned up by gel chromatography. They were separated on a 60 m fused silica capillary (conditions as for the volatile compounds listed in Table IV).

**Table V. Determination of Volatile Compounds in Tomato Paste and Tomato Flakes by Extraction with Methylene Chloride**

| Compound | Tomato paste (mg/kg DM) | Tomato flakes (mg/kg DM) | Sensory threshold values (mg/L $H_2O$) |
|---|---|---|---|
| Furfural | < 0.05 | 11.1 | 3.0 |
| 5-Methylfurfural | < 0.05 | 0.5 | 10* |
| 2-Acetylfuran | < 0.05 | 0.36 | 110* |
| Phenylacetaldehyde | 0.3 | 31.2 | $4 \times 10^{-3}$ |
| 2-Acetylpyrrole | < 0.05 | 2.0 | 200* |
| 2-Formylpyrrole | < 0.05 | 16.5 | |
| Furaneol | 0.42 | 8.2 | 0.03 |
| HMF | 14.0 | 249.0 | 200* |
| DHM | < 0.05 | 189.0 | 200* |

* determined in orange juice

The Maillard products furaneol, DHM and HMF could not be isolated by SDE (cf. Table IV) because of their low volatility; for the same reason the concentrations of 2-acetylpyrrole and 2-formylpyrrole isolated by SDE are lower than their concentrations found by extraction (Table V). According to Table V all Maillard products listed show a strong increase in the flakes; therefore in principle all these compounds could be used as chemical markers for recognizing chemical and/or sensory changes due to thermal processing of tomatoes. However, by comparing Table V with Table IV it becomes clear that big portions of furfural, 5-methylfurfural, 2-acetylfuran and phenylacetaldehyde are formed as artefacts during the SDE procedure. Nevertheless, the difference in concentrations between the tomato paste and the tomato flakes remains high.

**Conclusion**

By analytical determination of characteristic marker substances it is possible to trace the effects of thermal treatment during different processing steps on changes in product quality. In this way thermal processing steps can be controlled so that tolerable amounts of undesirable compounds are not exceeded.

In the case of tomato processing pyrrolidone-carboxylic acid (PCA) formed by cyclization of glutamine turned out to be the most sensitive marker for an early detection of heat treatment. PCA already appears during the break process. In tomato paste all the glutamine present has been converted to PCA as a result of heat treatment during the concentration process, whereas only relatively small amounts of hydroxymethylfurfural (HMF) could be detected.

Fructose-pyrrolidone-carboxylic acid (Fru-PCA) is the first Amadori compound which arises due to thermal processing. In tomato paste also small amounts of fructose-γ-aminobutyric acid (Fru-Gaba) and fructose-glutamic acid (Fru-Glu) could be detected. The most severe changes occur during the drying process. In drum-dried tomato flakes big portions of the amino acids present in tomato paste have been converted to the corresponding Amadori compounds which are precursors of flavor and off-flavor development. Furthermore, a series of volatile Maillard reaction products is formed as a consequence of thermal processing. Drum-drying of tomato paste causes a very strong increase in the concentrations of the Strecker aldehydes 2- and 3-methylbutanal as well as phenylacetaldehyde; they contribute to flavor changes because of their low sensory threshold values. They can be used as markers for undesirable sensory changes during heat processing. Moreover, a series of other Maillard products like furan and pyrrole derivatives as well as HMF and 5-hydroxy-5,6-dihydromaltol (DHM) show a very strong increase in concentrations as a result of the drying process.

## Literature Cited

1.    Leoni, C. et al. *Industria Conserve* **1979**, *54*, 199-204.
2.    Eichner, K.; Ciner-Doruk, M. *Z. Lebensm. Unters. Forsch.* **1979**, *168*, 360-367.
3.    Goose, P.G.; Binsted, R. *Tomato Paste;* Food Trade Press Ltd.: London, **1973**.
4.    dall'Aglio, G.; Carpi, G.; Versitano, A.; Palmieri, L. *Industria Conserve* **1985**, *60*, 187-192.
5.    Eichner, K.; Reutter, M.; Wittmann, R. In *Thermally Generated Flavors*; Parliment, Th.H.; Morello, M.J.; McGorrin, R.J., Eds.; ACS Symposium Series No. 543; American Chemical Society: Washington, DC, 1994; pp 42-54.
6.    Wittmann, R.; Eichner, K. *Z. Lebensm. Unters. Forsch.* **1989**, *188*, 212-220.
7.    Hendel, C.E.; Silveira, V.G.; Harrington, W.O. *Food Technol.* **1955**, *9*, 433-438.
8.    Rockland, L.B. *Analyt. Chem.* **1960**, *32*, 1375-1376.
9.    Reutter, M.; Eichner, K. *Z. Lebensm. Unters. Forsch.* **1989**, *188*, 28-35.
10.   Namiki, M. *Adv. Food Res.* **1988**, *32*, 115-184.
11.   Baltes, W. *Dtsch. Lebensm. Rdsch.* **1979**, *75* , 2-7.
12.   Baltes, W. *Lebensmittelchem. Gerichtl. Chem.* **1980**, *34*, 39-47.
13.   Danehy, J.P. *Adv. Food Res.* **1986**, *30*, 77-138.
14.   Sapers, G.M. *J. Food Sci.* **1970**, *35*, 731-733.
15.   Eichner, K. *Dtsch. Lebensm. Rdsch.* **1973**, *69*, 4-12.
16.   Baltes, W. *Lebensmittelchemie* **1993**, *47*, 9-14.
17.   Schönberg, R.; Moubacher, R. *Chem. Rev.* **1952**, *50*, 261-277.
18.   Piloty, M.; Baltes, W. *Z. Lebensm. Unters. Forsch.* **1979**, *168*, 368-373; 374-380.

# Chapter 5

# Chemical Markers for the Protein Quality of Heated and Stored Foods

H. F. Erbersdobler, J. Hartkopf, H. Kayser, and A. Ruttkat

Institute for Human Nutrition and Food Science, University of Kiel Düsternbrooker Weg 17, D–24105 Kiel, Germany

In heat treated or stored food products several amino acids are not fully available because of derivatization or crosslinking reactions. Since 30 years furosine is known as a useful indicator of early Maillard reaction which is applied in food science, nutrition and medical biochemistry. Recently more sensitive analytical methods for furosine determination are available which have again increased the attractivity of this important indicator. Lately, $N^\varepsilon$-carboxymethyllysine (CML) became available as another marker of special interest, because CML is a more useful indicator of the advanced heat damage by Maillard reaction than furosine. In addition, CML has the advantage to indicate reactions of lysine with ascorbic acid or ketoses such as fructose. Indicators for protein oxidation of sulfur amino acids are methionine sulfoxide and cysteic acid. An established marker for cross-linking reactions is lysinoalanine, which also indicates protein damages due to processing under alkaline conditions. Other markers formed as a consequence of alkaline treatment are D-amino acids.

Since about 50 years chemical markers for the protein quality of heated food are in use. For the first amino acid analysers were used in order to evaluate protein quality (*1*). However soon it was known that this procedure was not reliable in heat damaged food (*2*). Several amino acids proved to be more sensitive than others like lysine, cystine and tryptophan. Especially lysine was found to be a good marker for heat damage for it has a free ε-amino group able to react with other food ingredients like reducing sugars in the Maillard condensation or with other amino acids. Moreover it is abundant in many proteins and is additionally more easily to analyze than others. But soon it had to be recognized that in usually heat treated or stored food less destruction of lysine but more derivatisation or crosslinking occurs, a fact which was not considered by the usual lysine analysis.

One of the first proposals for the determination of available lysine was the derivatisation of the critical ε-amino group with fluoro dinitrobenzene (*3, 4, 5, 6*) which led to a worldwide application of this techniques. Later on many other derivatisation techniques for the determination of the "reactive lysine" were tested, proposed and applied (*7,8*).

However, to evaluate the intensity of the heating process it is possible either to measure losses of nutrients or to analyze the concentrations of new substances whose formation depends on different heating conditions. Since meanwhile more sensitive indicators are wanted, suited to measuring also a low technical impact on protein quality e.g. in UHT-treated milk, these reaction products became more iportant.

The first main intermediates formed during the Maillard reaction in the most common food items are the Amadori compounds with the fructoselysine moiety (fructoselysine, lactuloselysine or maltuloselysine) which are degraded during the acid hydrolysis of the protein, necessary for amino acid analysis. However they can be estimated by analyzing for furosine which is formed during hydrolysis with strong HCL (Fig. 1).

**Furosine and Other Markers for the Initial Stage of the Maillard Reaction**

Since the detection of furosine as a stable indicator of the Amadori products (*9, 10, 11, 12*) it was used as indicator of thermal damage in food science and medical biochemistry (see summarized in *13*). Recently its importance was enhanced by analytical improvements starting with the proposal of Resmimi et al. (*14*). Additionally the fact that now a pure and stable standard is commercially available led to further analytical activities (e.g. *15, 16*). Furosine has the disadvantage that it is formed from the Amadori products only at a rate of 30-40%. However, this recovery is reproducible if constant analytical conditions are

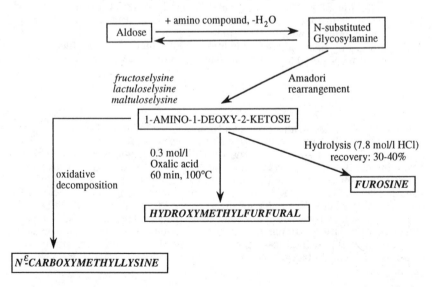

Figure 1. Various markers for the detection and quantification of Amadori products.

applied. Moreover furosine increases not linearly with increasing heat damage since the Amadori products are only intermediates which react to further compounds in the advanced and late Maillard reaction. In this way furosine is suited as parameter mainly in slighly of moderately heated products or after storage under favourable conditions repectively. In more severely heated products furosine values reach a plateau or even decrease.

On the other hand, furosine has the big advantage that it is a direct marker for a real existing reaction product of lysine, which has nutritional relevance. It represents the main reaction product of the initial stage of Maillard condensation and in this way a sector of heat damage, which is most interesting for heated milk and milk products.

A comparison of the furosine values from model experiments on a pilot plant working either by direct or indirect heating is given in figure 2. The data which are already published (*17*) are corrected according to the results with the meanwhile available pure furosine standard. The results show that there is a clear relationship between the severity of the heat treatment in terms of heating time and temperature and the furosine values of the milks.

In dairy products also hydroxymethylfurfural (HMF) and lactulose are common markers. For the determination of HMF, which also results from the Maillard condensation (*18*), precursors of browning products in milk are transformed to HMF after addition of oxalic acid and following heating (*19*). Principally, the HMF value of a milk can be used as an indicator for the heating process, but data from literature offer a wide range for this value which suspect that the HMF determination is insufficiently reproducible between laboratories. In particular, the level of HMF in untreated material, measured during the determination of "total HMF" and subtracted from the levels in treated milks, is a source of variation (*20*). However a comparison between the furosine and the HMF-method demonstrated the usefulness of the HMF-method as a rapid and simple measure of heat damage caused by the UHT process (*17*).

Also for HMF the correlation to heat damage is not linear while the third parameter of heat damage - lactulose - is linearly formed from lactose during heating or storage (*21, 22*). However the formation of lactulose changes depending also on secondary conditions of heat treatment like dry matter content or pH.

A comparison between the three parameters fromout results of a model experiment (*23*) is given in figure 3. The comparison shows that lactulose exhibits the most linear response to heating while furosine has a curvilinear characteristic. As can be seen in the more severely heated milk samples a fourth indicator, $N^{\varepsilon}$-Carboxy-methyllysine (CML), is appearing, although in very low concentrations.

## Crosslinking of proteins - Lysinoalanine (LAL)

Additionally to the above mentioned and widely used markers, lysinoalanine (LAL) was determined in model experiments and in several commercial products. LAL is formed throughout heat and/or alkali treatment of proteins by nucleophilic reaction of the lysyl-$\varepsilon$-amino-group with the activated double bond of dehydroalanine, which is formed by ß-elimination of cystine and phosphoserine in the peptide chain. Unlike furosine, LAL crosslinking creates not only a decrease in lysine but mainly of cyst(e)ine availability in the case of alkaline treatment.

As an example we examined the variance of the LAL values from 190 commercial samples collected from 45 different dairies in the western part of Germany. The normal range for LAL values of 0.5 - 5.8 mg/l was obtained as given in figure 4. Higher values revealed excessive heating and confirm other findings that thermal processes are often too severe.

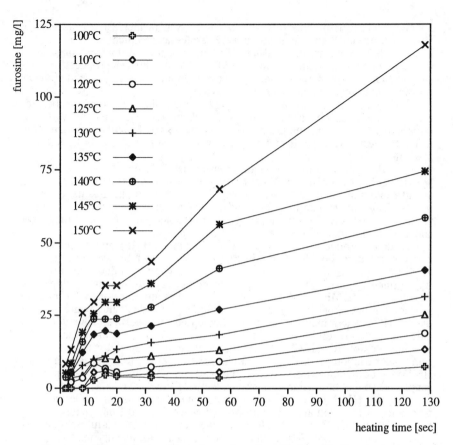

Figure 2. Comparison of furosine values in milk heated directly in an UHT pilot plant (recalculated from Dehn-Müller et al., 1989 and corrected according to the new furosine standard).

## N$^\varepsilon$-Carboxymethyllysine (CML) - A Marker for the Advanced Maillard Reaction

In the advanced and late "Maillard reaction" the furosine method is not valuable since the Amadori products lead to further compounds. Therefore the amount of available lysine in severely damaged foods is underestimated. In the case of fructose rich food (e.g. special diets for diabetics or products which are sweetened with honey or concentrated fruit juice) similar false conclusions are possible since glucoselysine which is formed from fructose during the "Maillard reaction" is not a precursor for furosine.

As an alternative method for the above mentioned food items we proved the formation of N$^\varepsilon$-carboxymethyllysine (CML) in a model system of several sugars and lysine (24) or casein, respectively. CML, an oxidative decomposition product of $\varepsilon$-fructoselysine was identified for the first time in 1985 (25) in biological material and later also in food items (26, 27). This new compound can be detected

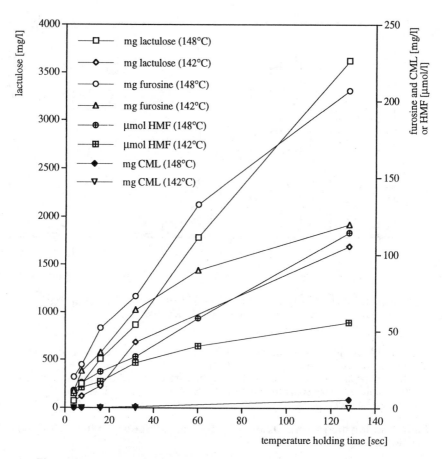

Figure 3. Relative values of Furosine, HMF, lactulose and CML after UHT-heating in a pilot plant (calculated from Hewedy et al., 1994)

by reversed-phase HPLC analysis after precolumn derivatization with o-phtaldialdehyde (*28*).

Figure 5 shows the formation of CML from various sugars and especially from fructose in a model system consisting of lysine and the respective sugars. The contents of CML were measured after heat treatment at 98°C for 3-24 h.

The results show for the first time that CML is formed by the reaction of lysine with fructose or sorbose. The CML values increase with increasing time of heating. After heating up to 6-12 h the reaction of lysine with ketoses leads to similar amounts of CML as the reaction with aldoses, whereas after longer heating significantly more CML was produced by the reaction with aldoses. The highest values were obtained by the reaction with galactose.

Latest results lead to a discussion about the mechanisms by which CML is formed (Fig. 6). Wells-Knecht et al. have proposed an alternative pathway which bases on the metal catalyzed oxidation of the sugar moiety to the dicarbonyl compound glyoxal, which itself reacts with lysine to CML under physiological conditions (*29*).

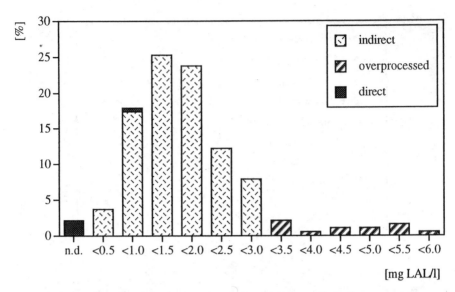

Figure 4. Variance of lysinoalanine (LAL) in 190 samples of commercial UHT milk (data recalculated from Dehn-Müller, 1991).

It is likely that the formation of CML out of the reaction of lysine with fructose is mainly due to the glyoxal pathway.

In our recent model studies on the mechanism of the CML formation under practical conditions we confirmed the existence of glyoxal as an intermediate from aldoses and ketoses.

**Indicators for the Protein Oxidation - Damage of Sulfur Amino Acids and Tryptophan**

An oxidation of food proteins leads partially to changes in the nutritional and functional properties and the sulfur amino acids methionine and cystein as well as histidine, tyrosine and tryptophan are susceptible to oxidation even under relatively mild conditions. Protein-bound forms of these amino acids can also be oxidized under conditions that sometimes prevail in stored food.

Methionine and other sulfur amino acids may be oxidized to the sulphoxide stage during food processing which has been pointed out by several authors in the past (*30, 31, 32*). Free tryptophan in food and simple systems oxidizes readly in the presence of light or peroxidizing lipids, participates in the Maillard reaction and is destroyed under the usual conditions for acid hydrolysis (*31, 32*). The loss of tryptophan through this oxidation is of great importance because it is an essential amino acid and the resulting products can contribute to off-flavours in irradiated foods and other undisired effects (*33*).

**Racemization of Amino Acids**

The conversion of L-amino acids in food proteins into D isomers generates nonutilizable forms of amino acids, creates peptide bonds resistant to proteolytic

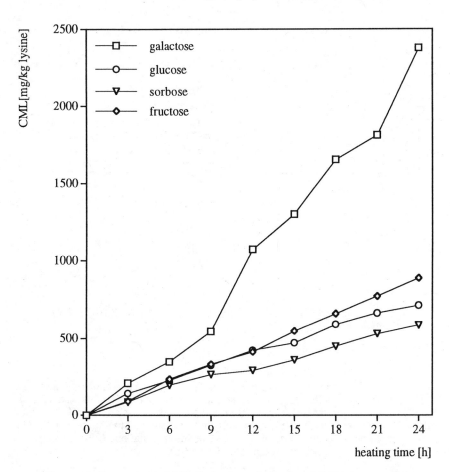

Figure 5. Formation of Nε-Carboxymethyllysine (CML) in a model system with lysine monohydrochloride and various reducing sugars after heating at 98°C (recalculated from Ruttkat and Erbersdobler, 1995).

enzymes, and forms unnatural amino acids that may be nutritionally antagonistic or even toxic (*34*). These razemisation reaction were found as a consequence of technological treatment of food, conservation and fermentation as well as cooking and backing of foods. Especially under alkaline conditions the reaction is thought to proceed by abstraction of the α-Proton from an amino acid like in the case of phoshorylserine and cysteine as β-elimination with dehydroalanine as an intermediate.

Schäfer et al. (*35*) studied the alkaline-induced racemization of several food proteins. For extruded casein as well as pretzel sticks no racemization could be detected. However, samples of gelatine showed high contents of D-amino acids. It is concluded that D-asparagine and D-serine are valuable markers for overprocessed foods and that their absence is a sign of "good manufacturing practice".

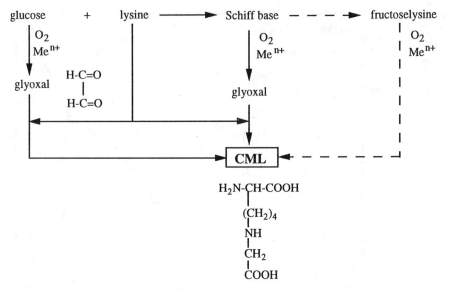

Figure 6. Proposed pathway for the formation of CML (Modified according to Wells-Knecht et al., 1995).

Some years ago consumers were disturbed by a report that noxious compounds including D-amino acids are produced in microwaved food (36). However, several authors found afterwards that there are no differences in the content of D-amino acids between the microwave heating and conventional cooking methods (37, 38, 39). Also in the concentrations of the above mentioned markers (furosine ect.) no microwave-specific changes in food were found (40, 41).

## References

1.  Mitchell, H.H.; Block, J. *J. Biol. Chem.* **1946**, 163, 599
2.  Oser, B.L. *J. Am. Diet. Assoc.* **1951**, 27, 396
3.  Carpenter, K.J.; Ellinger, G.M. *Biochem. J.* **1955**, 61, xi
4.  Carpenter, K.J. *Biochem. J.* **1960**, 77, 604
5.  Schober, R.; Prinz, I. *Milchwissenschaft* **1956**, 11, 466
6.  Erbersdobler, H. F; Zucker, H. *Z. Tierphysiol. Tierern. Futtermittelkde.* **1964**, 19, 244
7.  Hurrell, R.F.; Carpenter, K.J. *Br. J. Nutr.* **1974**, 32, 589
8.  Erbersdobler, H. F.; Anderson, T. R. In: *The Maillard-reaction in foods and nutrition*, eds. Waller, G. R.; Feather, M. S.; ACS Symp. Ser. 215; American Chemical Society: Washington D.C., 1983, pp. 419-427.
9.  Erbersdobler, H. F.; Zucker, H. *Milchwissenschaft* **1966**, 21, 564
10. Brüggemann, J.; Erbersdobler, H. F. *Z. Lebensmittel-Untersuchung und - Forschung* **1968**, 137, 137
11. Heyns, K.; Heukeshoven, J.; Brose, K.-H. *Angew. Chemie* **1968**, 80, 627
12. Finot, P. A.; Bricout, J.; Viani, R.; Mauron, J. *Experientia* **1968**, 24, 1097

13. Erbersdobler, H. F In: *Amino carbonyl reactions in food and biological systems*; eds. Fujimaki, M.; Namiki, M.; Kato, H.; Elsevier: Amsterdam and Kodansha, Tokio, 1986, pp. 481-491.
14. Resmimi, P.; Pellegrino, L.; Batelli, G. *Ital. J. Food Sci*. **1990**, 3, 173
15. Hartkopf, J.; Erbersdobler, H. F. *J. Chromat*. **1993**, 635, 151
16. Henle, T.; Zehetner, G.; Klostermeyer, H. Z. *Lebensmittel-Untersuchung und -Forschung* **1995**, 200, 235
17. Dehn-Müller, B. In: *Schriftenreihe des Instituts für Humanernährung und Lebensmittelkunde*; ed. Erbersdobler, H. F.; Kiel, 1989, Vol. 5
18. Keeney, M.; Basette, R. *J. Dairy Sci*. **1959**, 6, 945
19. Konietzko, M. *Agr. Diss*., **1981**
20. Burton, H. *J. Dairy Res*. **1984**, 51, 341
21. Geier, H.; Klostermeyer, H. *Milchwissenschaft* **1983**, 38, 475
22. Andrews, G. *IDF Bulletin* 238 **1989**, 52
23. Hewedy, M. M.; Kiesner, C.; Meissner, K.; Hartkopf, J.; Erbersdobler, H. F. *J. Dairy Res*. **1994**, 61, 305
24. Ruttkat, A.; Erbersdobler, H. F. *J. Sci. Food Agri*. **1995**, 68, 261
25. Ahmed, M. U.; Thorpe; S. R.; Baynes, J. W. *J. Biol. Chem*. **1986**, 261, 4889
26. Büser, W.; Erbersdobler, H. F. *Milchwissenschaft* **1986**, 41, 780
27. Büser, W.; Erbersdobler, H. F.; Liardon, R. *J. Chromatogr*. **1987**, 387, 515
28. Hartkopf, J.; Pahlke, C.; Lüdemann, G.; Erbersdobler, H. F. *J. Chromat. A* **1994**, 672, 242
29. Wells-Knecht, K. J.; Zyzak, D. V.; Litchfield, J. E., Thorpe, S. R.; Baynes, J. W. *Biochemistry* **1995**, 34, 3702
30. Gjøen, A. U.; Njaa, L. R. *Br. J. Nutr*. **1977**, 37, 93
31. Nielsen, H. K., Finot, P. A., Hurrell, R. F. *Br. J. Nutr*. **1985**, 53, 75
32. Nielsen, H. K., De Weck, D., Finot, P. A., Liardon, R., Hurrell, R. F. *Br. J. Nutr*. **1985**, 53, 281
33. Krogull, M. K.; Fennema, O. *J. Agric. Food Chem*. **1987**, 35, 66
34. Masters, P. M.; Friedman, M. In: *Chemical Deterioration of Proteins*, eds. Whitaker, J. R.; Fujimaki, M., American Chemical Society Symposium Series, Vol. 123; American Chemical Society: Washington D.C., **1980**, p. 165
35. Schäfer, S.; Bahnmüller, D.; Hausch, M.; Brückner, H. *Ernähr.-Umsch*. **1987**, 34, 84
36. Lubec G., Wolf G., Bartosch S. *Lancet II* 1989, Dec. 9., 1392-1393
37. Fay, L., Richli, U., Liardon, R. *J.Agric. Food Chem*. 1991, **39**, 1857-1859
38. Marchelli, R., Dossena, A., Palla, G., Audhuy-Peaudecerf, M., Lefeuvre, S., Carnevali, P. and Freddi, M. *J. Sci. Food Agric*. 1992, **59**, 123-126
39. Fritz, P., Dehne, L.I., Zagon, J. und Bögl, K,W. *Z. Ernährungswiss*. 1992, **31**, 219-224
40. Buhl, K., Smit, E. und Erbersdobler, H.F. *Lebensmittelchemie*, 1993, **47**, 113-114
41. Meißner, K. and Erbersdobler, H.F. *J. Sci. Food Agric*. 1996, **70**, inpress

Chapter 6

# Principles and Applications of Chemical Markers of Sterility in High-Temperature– Short-Time Processing of Particulate Foods

Hie-Joon Kim[1], Irwin A. Taub[1], Yang-Mun Choi[2], and Anuradha Prakash[3,4]

[1]U.S. Army Natick Research, Development and Engineering Center, Kansas Street, Natick, MA 01760–5018
[2]Institute of Biotechnology, Korea University, 1 Anam-dong, Sungbuk-ku, Seoul 136–701, Korea
[3]Department of Food Science and Technology, Ohio State University, Columbus, OH 43210

Continuous sterilization and aseptic packaging technologies have a great deal of potential to produce shelf-stable foods in convenient packages. A direct measurement of time-temperature history within food particulates is not practical in continuous, high temperature/short time (HTST) processes. The yield of thermally produced compounds offers an alternative as a time temperature integrator and as a chemical marker of sterility. One such a compound, 2,3-dihydro-3,5-dihydroxy-6-methyl-4(H)-pyran-4-one (M-1), is formed at sterilizing temperatures from D-glucose or D-fructose and amines through 2,3-enolization under weakly acidic or neutral conditions. Another marker, 4-hydroxy-5-methyl-3(2H)-furanone (M-2), is formed similarly from D-ribose or D-ribose-5-phosphate. Application of these compounds to mapping lethality distribution within food particulates in two volumetric heating processes, ohmic heating and microwave sterilization, is demonstrated.

Conventional thermal processing, such as retorting, relies on heat transfer from the surrounding heat source, often through a liquid medium, to the center of particulate foods. Therefore, when producing shelf-stable foods, a certain amount of overprocessing takes place by the time commercial sterility is achieved at the cold spot of the food particulates. Such overprocessing could be avoided if the particulates are sterilized by heat generation throughout the volume.

[4]Current address: Department of Food Science and Nutrition, Chapman University, Orange, CA 92666

Ohmic heating and microwave sterilization are two volumetric heating technologies available to food processors. In ohmic heating, the electrical conductivities of the fluid and the particulates are important parameters (*1,2*). In microwave sterilization, heat generation depends on the dielectric loss factor of the food materials (*3,4*). For industrial applications, both ohmic and microwave processes are carried out in a continuous mode. In ohmic heating, foods are continuously pumped through sets of electrodes under high voltage, holding tubes, and cooling tubes, and then aseptically packaged (*5*). In microwave processing, prepackaged foods are sterilized, under high pressure, with microwaves from magnetrons above and below the foods moving on a conveyer belt (*6*).

In either case, the time-temperature measurement within the moving food particulates is difficult, and consequently assuring commercial sterility without overprocessing is not a straightforward matter. In this paper, we will discuss how thermally produced compounds can be used as chemical markers of sterility in ohmic heating and microwave sterilization.

## Selection of the Chemical Markers

**Destruction vs. Formation.** When looking for chemical markers of sterility, one is tempted to look for compounds that are destroyed at sterilizing temperatures for the simple reason that the chemical identity and the assay method is already known to the investigator. Several examples were listed by Kim and Taub (*7*). This approach has a limitation, because a typical chemical reaction in foods is much slower than bacterial destruction at high temperatures and one has to be able to measure a small loss of the compound. For example, the D-value (time required to reduce the concentration by 90%) for destruction of thiamin is 244 min at 122°C (*8*). The D-value for destruction of *B. stearothermophilus* is about 1 min at the same temperature. The D-value and k, the rate constant for a first-order reaction, are related by eq. (1).

$$k = 2.303/D \qquad (1)$$

Thus, the rate constant for destruction of thiamin is 0.0094 min$^{-1}$ at 122°C. For commercial sterility, 5-7 min heating at 121°C is usually required (*9*). After 5 min heating at 122°C, e$^{-kt}$ equals 0.954 and only 5% loss of thiamin will take place, which is difficult to measure accurately. On the other hand, some reactions such as enzyme inactivation are too fast at sterilization temperatures and would be useful only as markers for pasteurization.

However, if one were to turn attention to the products of such slow reactions, accurate determination becomes much easier, because one starts with a zero baseline. The product (marker) concentration approaching a limiting value exponentially can be expressed as follows:

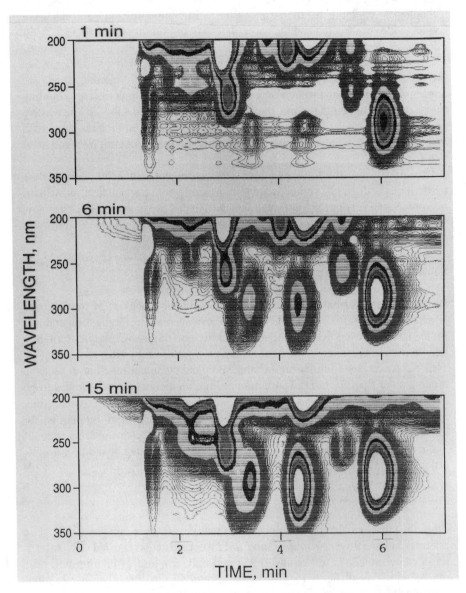

Figure 1   Contour diagram for water-extractable compounds in beef heated for 1, 6, and 15 min at 121°C. The x-axis is chromatographic retention time, the y-axis is uv wavelength, and the z-axis is absorbance.

$$M(t)/M_\infty = 1 - e^{-kt} \tag{2}$$

For kt $\ll$1, $e^{-kt}$ can be approximated as 1 - kt and eq. (2) becomes

$$M(t)/M_\infty = kt \tag{3}$$

which indicates that the marker concentration is directly proportional to the heating time at a given temperature. Particularly interesting possibilities exist in the case where two markers are formed with different rates and activation energies (*10*).

**Ease of Detection and Stability.** Numerous compounds are thermally produced in foods, but not all are suitable as chemical markers of sterility. Some of the compounds that need to be ruled out include volatiles and unstable intermediates that rapidly undergo subsequent reactions. Preferably, the marker compound should be easily extracted with an aqueous solvent and easily determined without many additional operations. The marker should also be stable during analysis. In situ analysis would be ideal; however, accurate quantitation by simple in situ methods, such as surface fluorescence or near infrared measurements, is questionable.

Figure 1 shows contour diagrams of spectrochromatograms of water-soluble compounds from beef heated for 1, 6, and 15 min at 121°C using pressurized steam. The three-dimensional spectrochromatogram was obtained using anion exclusion chromatographic separation and photodiode array detection (*7*). It is clear that the compound with elution time of 5.8 min and absorption maximum of 285 nm (M-2)(*7*) is formed quite rapidly and approaches a limiting value after 15 min heating. The formation of another compound with elution time of 4.2 min and absorption maximum of 298 nm (M-1)(*7*) is slower and still ongoing after 15 min. These compounds are easily extracted with water and determined by liquid chromatography (*7*). Spectrophotometric detection at a fixed uv wavelength (285 or 298 nm) could be used for simultaneous determination of both M-1 and M-2 without any interference. The heated sample or the extract could be frozen and stored for several days without affecting the analysis.

**Identification.** Purification and identification of M-1 and M-2 as 2,3-dihydro-3,5-dihydroxy-6-methyl-4(H)-pyran-4-one and 4-hydroxy-5-methyl-3(2H)-furanone, respectively, by mass spectrometry have been published (*7,11*). Analysis of different types of foods heated similarly at sterilizing temperatures revealed that M-1 is formed in meats and vegetables and M-2 is formed in meats only. Another compound, 5-hydroxymethylfurfural, appears to be a useful marker in heating fruits and fruit juices (M-3)(*7*).

## Earlier Work and Precursor-Marker Relationship

Both 2,3-dihydro-3,5-dihydroxy-6-methyl-4(H)-pyran-4-one (referred to as M-1 for convenience) and 4-hydroxy-5-methyl-3(2H)-furanone (referred to as M-2) have been known for almost 30 years. They were observed in heated foods and synthesized in simple model systems. We were interested in identifying the natural precursors of M-1 and M-2 in real foods.

**M-1.** Formation of M-1 was first noticed by Shaw et al. in 1967 in acid-catalyzed dehydration of D-fructose (*12*). In fact, they identified the compound as 4-hydroxy-2-(hydroxymethyl)-5-methyl-3(2H)-furanone. In the same year, the same authors observed the compound in stored dehydrated orange powder (*13*), which represents the first detection of M-1 from real foods. In 1968, Severin and Seilmeier reported a new compound formed from D-glucose, acetic acid, and methylamine and assigned an incorrect structure (*14*). The correct structure assignment was reported by Mills et al. in 1970 (*15*). In the following year, Shaw et al. confirmed the correct structure (*16*).

In 1976, Ledl et al. demonstrated that M-1 is formed in heated carrots, onions, tomatoes, cabbage and meat as well as in caramelized sugars and bread crust (*17*). Takei detected M-1 as an aroma compound in roasted sesame seed (*18*). In 1990, Nishibori and Kawakishi noted that M-1 is a major flavor component in baked cookies (*19*). They also reported that reaction between fructose and protein yield more M-1 than that between glucose and protein after 10 min baking at 150°C. They extended this work to fructose and β-alanine and again observed that a slightly higher yield of M-1 is obtained with fructose than with glucose (*20*).

M-1 was also produced from glucose and piperidine (*21*), lactose and lysine (*22*), glucose and proline (*23*), 1-deoxy-D-erythro-2,3-hexodiulose and piperidine (*24*), and glucose and propylamine (*25*). In 1976, Mills and Hodge showed that 1-deoxy-(L-proline)-D-fructose is converted to M-1 upon pyrolysis (*26*). In 1987, Njoroge et al. detected the formation of M-1 from glucose and neopentylamine under physiological conditions (*27*).

From these results, it appears that both gluocse and fructose will react with proteins and amino acids in foods to form M-1. In mammalian muscle after rigor mortis, approximately 0.17% is glucose-6-phosphate and 0.01% is glucose by wet weight (*28*). Ribose-6-phosphate was implicated in the formation of M-2 (*29-31*). We performed spiking experiments to determine whether glucose-6-phosphate or glucose is the natural precursor of M-1 in meats. One % D-glucose, D-glucose-6-phosphate, D-fructose, or D-ribose was added to a meat extract, and the time course of M-1 formation in the meat extract at 121°C was monitored up to 100 min. Figure 2 shows that the control meat extract and the ribose-added meat extract yield the same amount of M-1 from the precursors already present in the meat. Thus ribose served as another control. Addition of fructose increased the M-1 yield by approximately 70% over the control. Addition of the same amount of glucose increased the M-1 yield approximately three-fold. On the other hand, glucose-6-phosphate seemed to slightly decrease the M-1 yield. Glucose-6-phosphate might compete with glucose for reaction with the amines, but does not lead to the formation of M-1. It is not clear why fructose shows higher reactivity than glucose toward M-1 formation in baking and the reverse is true in heating meat. The concentration of fructose is much lower than that of glucose in meats; therefore, we believe that glucose, not glucose-6-phosphate, is the natural precursor of M-1 in meats. Fructose might play a more important role in M-1 formation in vegetables.

**M-2.** Severin and Seilmeier first synthesized M-2 from pentoses and primary amine salts in 1967 (*32*). In the following year, Peer et al. synthesized M-2 from D-xylose or D-ribose and secondary amine salts (*33*), and Peer and van den Ouweland synthesized M-2 similarly from D-ribose-5-phosphate (*29*). In the same year, Tonsbeek et al. published identification of M-2 from beef broth (*34*). Subsequently, Tonsbeek et al. identified ribose-5-phosphate and pyrrolidone carboxylic acid/taurine as natural precursors of M-2 in beef (*30*).

In 1974, Anderegg and Neukom showed that M-2 can be formed from purinnucleosides and -nucleotides as well as from pentoses upon heating without amines (*31*). Hicks et al. prepared M-2 from D-glucuronic acid and amine (*35*). In 1979, Nunomura et al. identified M-2 as an important flavor component in soy sauce, which is rich in pentoses such as arabinose, ribose, and xylose as well as amino acids (*36*). In 1980, Honkanen et al. identified M-2 in fresh wild raspberries (*37*). Knowles et al. observed M-2 resulting from addition of spinach chloroplast ribosephosphate isomerase to ribose-5-phosphate (*38*). Nursten and O'Reilly produced M-2 from glycine and xylose (*39*). Idstein and Schreier identified M-2 from guava fruit (*40*). M-2 was also identified from roasted sesame seed (*18*).

Above results indicate that M-2 can be formed from pentoses and their phosphates (*29,31,38*). Results of our spiking experiment (Figure 3) also show that both ribose and ribose-5-phosphate participate in the formation of M-2 in meats. It appears that the phosphate group is readily hydrolyzed from ribose-5-phosphate and the ribose participates in the formation of M-2, whereas glucose-6-phosphate is not hydrolyzed and does not participate in the formation of M-1. The relative concentrations of ribose, ribose-5-phosphate, and other pentoses in meats are not well known. Presumably, both ribose and ribose-5-phosphate are the primary precursors of M-2 in meats.

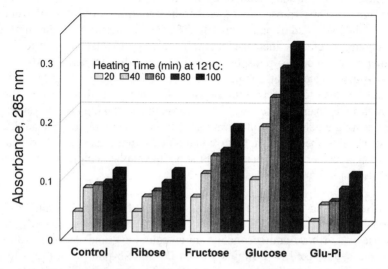

Figure 2   Results of spiking experiment showing change in M-1 yield upon heating beef extract with added ribose, fructose, glucose, and glucose-6-phosphate.

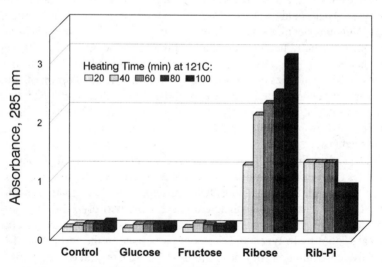

Figure 3   Results of spiking experiment showing change in M-2 yield upon heating beef extract with added glucose, fructose, ribose, and ribose-5-phosphate.

## Mechanism of the Marker Formation

In 1964, Anet described a dehydration pathway of Amadori compounds involving enolization and formation of dicarbonyl intermediates (*41*). It is believed that, in general, 1,2-enolization is favored in strongly acidic media, where the nitrogen atom of the Amadori compound is protonated, and subsequent dehydration leads to 2-furaldehyde from pentoses and 5-hydroxymethylfurfural from hexoses. On the other hand, under weakly acidic or alkaline conditions 2,3-enolization is favored leading to furanones and pyranones. In fact, Hicks and Feather showed that Amadori compounds form 2-furaldehyde in 2 N sulfuric acid and M-2 at pH 7 (*42*). Tressl et al. also showed that 5-hydroxymethylfurfural is a predominant product below pH 4 and M-1 is a major product above pH 5 from the proline/glucose model system (*43*). Feather stated that production of furaldehyde indicates a 1,2-enolization pathway and furanone a 2,3-enolization pathway (*44*). The 1,2- and 2,3-dicarbonyl intermediates are formed during dehydration of sugars (*12*) as well as of Amadori compounds.

We tested this reaction scheme in meat. Ten percent D-glucose was added to a beef extract and the pH was adjusted with a 6 N hydrochloric acid or sodium hydroxide solution. Figure 4 shows the pH-dependence of M-1 and M-3 after 15 min heating at 121°C. As expected, 1,2-enolization and M-3 formation predominates below pH 4. Above pH 5 there is little M-3 formation, which explains why we never observed M-3 from heated meats (pH 5.4). There is a sharp decline in the M-1 yield above pH 10. This observation is consistent with the base-catalyzed fructose dehydration Shaw et al. investigated at pH 11.5 (*45*). They reported that M-1 was not formed at pH 11.5; however, if the alkalinity was not constantly maintained, M-1 was formed as the pH decreased. Similarly, when 1% D-ribose was added to the beef extract, formation of 2-furaldehyde dominated at low pH and M-2 formation became more impotant at pH above 4.5, which again explains why 2-furaldehyde was never observed in heated meats. The general reaction pathways for formation of the markers are summarized in Figure 5.

## Applications to Ohmic Heating and Microwave Sterilization

In general, a given chemical marker concentration can be arrived at through many different time-temperature histories. The possibilities are drastically reduced when two markers are used. It may even be possible to determine a unique time-temperature history based on computer simulations if a practical range of time and temperature were given for the yield of two markers for which the reaction rate constants and activation energies are known. We are currently investigating this interesting aspect of the chemical marker application. In this paper, we will demonstrate how the markers, with certain limitations, can be used to provide key information in high temperature/short time (HTST) processing.

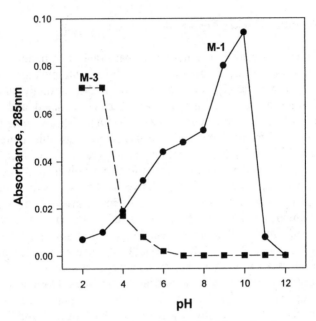

Figure 4    pH-dependence of M-1 and M-3 yield from beef extract with 1% glucose added.

**Ohmic Heating.**    In ohmic heating, foods are sterilized by electroresistive heat generation throughout the volume. A key advantage of ohmic heating is the possibility of faster heating in the particulates than in the fluid (*2,46-47*). If a range of product and process parameters (such as the fluid and particulate electrical and thermal conductivites, solids content, applied voltage, and flow rate) can be selected that will assure faster heating of the particulates, the fluid temperature can be measured and used to guarantee sterility of the entire food product. The chemical markers turned out to be extremely useful for demonstrating faster heating of the particulate center than the particulate surface or the fluid in a continuous ohmic heating system, where a direct temperature measurement within moving food particulates is nearly impossible.

Figure 6 shows that the concentrations of both chemical markers, M-1 and M-2, are higher at the center than near the surface of a meatball ohmically processed using a 5 kW system (*48*). At a given temperature, there is a linear relationship between log-reduction in bacterial population and the marker yield (*49*). Clearly, a higher lethality was achieved at the particulate center. Results in Table I, summarizing bacterial destruction predicted from the chemical marker yield and observed directly, also show higher lethality at the center at three operating temperatures.

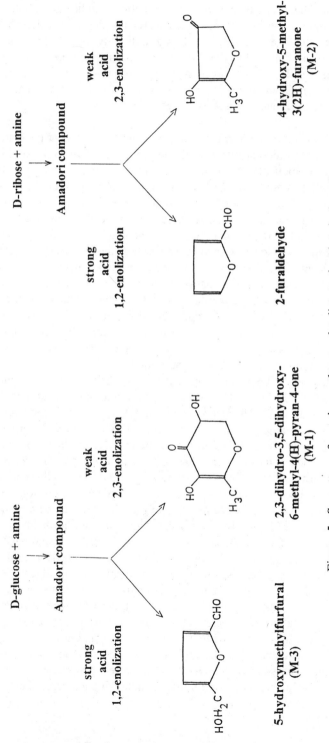

Figure 5 Summary of reaction pathways leading to the chemical marker formation.

This remarkable inversion in temperature is believed to be due to both faster heating and lower specific heat of the particulate than of the fluid. Faster heating of the particulates was also demonstrated indirectly by observation of a temperature rise in the fluid as the fluid-particulate mixture flows through the holding tubes of the ohmic processing system (H.-J. Kim et al., 1995, unpublished data). Such volumetric heating makes ohmic heating an attractive technology for producing high quality, shelf-stable foods (50).

**Microwave Sterilization.** In microwave heating of foods, heat generation is primarily due to the dipolar water molecules trying to align to the oscillating electrical field component of the microwave. The microwave energy is needed to break the hydrogen bonds in water to allow the dipoles to align. As the water molecules reform hydrogen bonds, the stored energy is dissipated as thermal energy (51). Several parameters, such as dielectric properties, shape, and size of the food, play important roles in microwave heating in generating the temperature profile in the food. At a low electrolyte concentration, the microwave penetration depth is significant and center heating can occur through a focusing effect when the object has a cylindrical or a spherical shape (4,52,53). As the electrolyte concentration is increased, the conductivity loss becomes important and more microwave power is dissipated at the surface creating a sharp temperature gradient (surface heating)(4,54). These observations imply that in principle it should be possible to select the salt concentration and the shape of the food that would provide a uniform heating throughout the food.

To test this idea, cylindrically shaped ham (3 cm diameter, 6 cm height) was prepared with 0.5, 1.0, 2.6, and 3.5% salt in addition to other usual ingredients. Microwave heating was performed in a cylindrical, pressurized container using a CEM microwave sample preparation system (MDS-2000, CEM Corporation, Matthews, NC). After heating for 2 min after the temperature of 121°C has been reached at the center of the cylinder (come-up time about 45 sec), the ham sample was cut into three equal sections along the height and the middle section was again cut radially into three portions (outer, middle, and the core portion). Figure 7 shows the M-2 yield in different portions of the ham. For the ham with 0.5% salt, the core showed higher M-2 yield than the outer ring probably due to a focusing effect. At 2.6 and 3.5% salt concentration, the outer ring showed much higher M-2 yield than the core. Clearly, the penetration depth is decreased as salt concentration is increased as observed previously by Mudgett (4) and Anantheswaran (54). It is interesting that the marker yield at the core is higher for the ham with 3.5% salt than that with 2.6% salt, even though the penetration depth should be smaller. It appears that, for the microwave that has penetrated, the conductivity loss is greater with the higher salt concentration.

At 1.0% salt concentration, the M-2 yield showed no gradient suggesting a uniform heating. This situation represents a balance between surface heating and focusing. Obviously, the salt concentration for achieving a uniform heating will

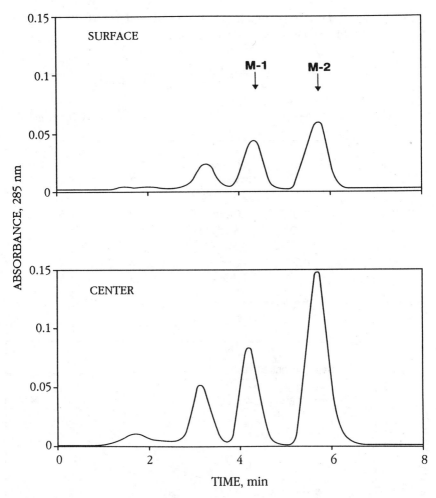

Figure 6   Chromatograms showing higher yield of M-1 and M-2 at the center than at the surface of an ohmically processed meatball. Reproduced with permission from Ref. 48.

depend on the size of the food.  This example illustrates that the marker yield can be used to map the temperature distribution within a solid food and to optimize the product formula for uniform heating.  The technique is also expected to be useful in verifying the thermal contour in microwave sterilized meals in various package geometries (*55*).

CHEMICAL MARKERS FOR PROCESSED AND STORED FOODS

**Table I. Bacterial Destruction in Ohmically Heated Meatballs Predicted from M-1 Yield and Observed Microbiologically**

| Temperature (°C) | Flow rate (L/min) | Location | log(N$_o$/N) | |
|---|---|---|---|---|
| | | | Predicted | Observed |
| 125 | 1.1 | Center | 7.8 | > 5 |
| | | Surface | 3.2 | 4.1 |
| 128 | 0.9 | Center | 13.9 | > 5 |
| | | Surface | 5.0 | > 5 |
| 132 | 0.9 | Center | 23.8 | > 5 |
| | | Surface | 13.5 | > 5 |

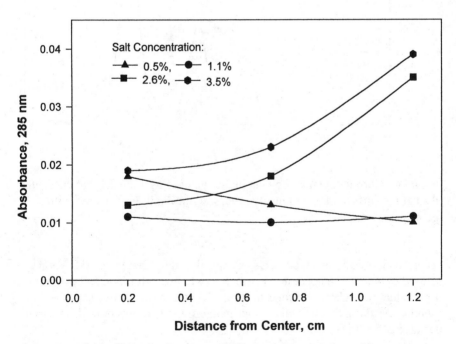

Figure 7. Yield of M-2 in the core, middle, and outer portion of ham containing different amounts of salt heated at 121°C for 5 min with microwave.

## Conclusion

The reported chemical markers are useful markers of sterility, which is an important quality index in shelf-stable foods. The use of the markers to map lethality distribution in particulate foods has been demonstrated. The markers can be used for validating and optimizing new thermal processing technologies such as ohmic heating and microwave sterilization.

## Literature Cited

1. Biss, C.H.; Coombes, S.A.; Skudder, P.J. In *Process Engineering in the Food Industry*; Field, R.W.; Howell, J.A., Eds.; Elsevier: London, 1989; pp 17-25.
2. Sastry, S.K.; Palaniappan, S. *Food Technol.* **1992**, 46(12), 64-67.
3. Ohlsson, Th.; Bengtsson, N.E. *J. Microwave Power.* **1975**, 10(1), 93-108.
4. Mudgett, R.E. *Food Technol.* **1986**, 40(6), 84-93, 98.
5. Parrott, D.L. *Food Technol.* **1992**, 46(12), 68-72.
6. Harlfinger, L. *Food Technol.* **1992**, 46(12), 57-61.
7. Kim, H.-J.; Taub, I.A. *Food Technol.* **1993**, 47(1), 91-97, 99.
8. Mulley, E.A.; Stumbo, C.R.; Hunting, W.M. *J. Food Sci.* **1975**, 40, 993-996.
9. Dignan, D.M.; Berry, M.R.; Pflug, I.J.; Gardine, T.D. *Food Technol.* **1989**, 43(3), 118-121, 131.
10. Ross, E. *J. Food Process Eng.* **1993**, 16, 247-270.
11. Kim, H.-J.; Ball, D.; Giles, J.; White, F. *J. Agric. Food Chem.* **1994**, 42, 2812-2816.
12. Shaw, P.E.; Tatum, J.H.; Berry, R.E. *Carbohyd. Res.* **1967**, 5, 266-273.
13. Tatum, J.H.; Shaw, P.E.; Berry, R.E. *J. Agric. Food Chem.* **1967**, 15(5), 773-775.
14. Severin, Th.; Seilmeier, W. *Z. Lebensm. Unters. Forsch.* **1968**, 137, 4-6.
15. Mills, F.D.; Weisleder, D.; Hodge, J.E. *Tetrahedron Letters,* **1970**, 15, 1243-1246.
16. Shaw, P.E.; Tatum, J.H.; Berry, R.E. *Carbohyd. Res.* **1971**, 16, 207-211.
17. Ledl, F.; Schnell, W.; Severin, Th. *Z. Lebensm. Unters. Forsch.* **1976**, 160, 367-370.
18. Takei, Y. *Nihon Kasei Gakkai Shi,* **1988**, 39(8), 803-815.
19. Nishibori, S.; Kawakishi, S. *J. Food Sci.* **1990**, 55(2), 409-412.
20. Nishibori, S.; Kawakishi, S. *J. Agric. Food Chem.* **1994**, 42(5), 1080-1084.
21. van den Ouweland, G.A.M.; Peer, H.G. *RECUEIL,* **1970**, 89, 750-754.
22. Ferretti, A.; Flanagan, V.P. *J. Agric. Food Chem.* **1973**, 21(1), 35-37.

23. Shigematsu, H.; Shibata, S.; Kurata, T.; Kato, H.; Fujimaki, M. *J. Agric. Food Chem.* **1975**, 23(2), 233-237.
24. Fisher, B.E.; Sinclair, H.B.; Goodwin, J.C. *Carbohyd. Res.* **1983**, 116, 209-215.
25. Knerr, T.; Severin, T. *Z. Lebensm. Unters. Forsch.* **1993**, 196, 366-369.
26. Mills, F.D.; Hodge, J.E. *Carbohyd. Res.* **1976**, 51, 9-21.
27. Njoroge, F.G.; Sayre, L.M.; Monnier, V.M. *Carbohyd. Res.* **1987**, 167, 211-220.
28. Lawrie, R.A. *Meat Science, 2nd Edition*; Pergamon Press; New York, NY, **1974**, p. 71.
29. Peer, H.G.; van den Ouweland, G.A.M. *RECUEIL*, **1968**, 87, 1017-1020.
30. Tonsbeek, C.H.T.; Koenders, E.B.; van der Zijden; Losekoot, J.A. *J. Agric. Food Chem.* **1969**, 17(2), 397-400.
31. Anderegg, P.; Neukom, H. *Lebensmitt. Wiss. Technol.* **1974**, 7(4), 239-241.
32. Severin, Th.; Seilmeier, W. *Z. Lebensm. Unters. Forsch.* **1967**, 134, 230-233.
33. Peer, H.G.; van den Ouweland, G.A.M.; de Groot, C.N. *RECUEIL*, **1968**, 87, 1011-1016.
34. Tonsbeek, C.H.T.; Plancken, A.J.; Weerdhof, T.v.d. *J. Agric. Food Chem.* **1968**, 16(6), 1016-1021.
35. Hicks, K.B.; Harris, D.W.; Feather, M.S.; Loeppky, R.N. *J. Agric. Food Chem.* **1974**, 22(4), 724-725.
36. Nunomura, N.; Sasaki, M.; Yokotsuka, T. *Agr. Biol. Chem.* **1979**, 43(6), 1361-1363.
37. Honkanen, E.; Pyysalo, T.; Hirvi, T. *Z. Lebensm. Unters. Forsch.* **1980**, 180-182.
38. Knowles, F.C.; Chanley, J.D.; Pon, N.G. *Arch. Biochem. Biophys.* **1980**, 202(1), 106-115.
39. Nursten, H.E.; O'Reilley, R. In *The Maillard Reaction in Foods and Nutrition;* Waller, G.; Feather, M.S., Eds.; ACS Symposium Series No. 215, American Chemical Society: Washington, DC, 1983, pp 103-121.
40. Idstein, H.; Schreier, P. *J. Agric. Food Chem.* **1985**, 33(1), 138-143.
41. Anet, E.F.L.J. *Adv. Carbohyd. Chem.* **1964**, 19, 181-218.
42. Hicks, K.B.; Feather, M.S. *J. Agric. Food Chem.* **1975**, 23(5), 957-960.
43. Tressl, R.; Helak, B.; Martin, N.; Kersten, E. In *Thermal Generation of Aromas;* Parliament, T.H.; McGorrin, R.J.; Ho, C.-T., Eds.; ACS Symposium Series No. 409, American Chemical Society: Washington, D.C, 1989, pp 156-171.
44. Feather, M.S. In *Maillard Reactions in Food*; Eriksson, C.; Ed.; Prog. in Food and Nutrition Science, Vol 5, Pergamon Press: NY, 1981, pp 37-45.

45. Shaw, P.E.; Tatum, J.H.; Berry, R.E. *J. Agric. Food Chem.* **1968**, 16(6), 979-982.
46. de Alwis, A.A.P.; Halden, K.; Fryer, P.J. *Chem. Eng. Res. Des.* **1989**, 67, 159-168.
47. de Alwis, A.A.P.; Fryer, P.J. *Chem. Eng. Sci.* **1990,** 45, 1547-1559.
48  Kim, H.-J.; Choi, Y.-M.; Yang, A.P.P.; Yang, T.C.S.; Taub, I.A.; Giles, J.; Ditusa, C.; Chall, S.; Zoltai, P. *J. Food Process. Preserv.* **1996**, 20, 41-58.
49. Ramaswamy, H.S.; Awuah, G.B.; Kim, H.-J.; Choi, Y.-M. *Activities Report/Minutes of Work Groups-R&D Associates* **1995**, 47(1), 216-225.
50. Skudder, P.J. In *Aseptic Processing and Packaging of Particulate Foods;* Willhoft, E.M.A., Ed.; Blackie Academic & Professional: London, 1993; pp. 74-89.
51. Walker, J. *Scientific American* **1987**, pp 134-138.
52. Ohlsson, T.; Risman, P.O. *J. Microwave Power* **1978**, 13(4), 303-310.
53. Padua, G.W. *J. Food Sci.* **1993**, 58(6), 1426-1428.
54. Anantheswaran, R.C.; Liu, L. *J. Microwave Power & Electromag. Energy* **1994**, 29(2), 119-126.
55. Coles, R.E. In *Aseptic Processing and Packaging of Particulate Foods;* Willhoft, E.M.A., Ed.; Blackie Academic & Professional: London, 1993; pp. 112-147.

Chapter 7

# The Effect of Processing on the Chiral Aroma Compounds in Cherries (*Prunus avium* L.)

Kieran Pierce[1], Donald S. Mottram[2], and Brian D. Baigrie[1]

[1]Reading Scientific Services Limited, Lord Zuckerman Research Centre, University of Reading, Whiteknights, Reading RG6 6LA, United Kingdom
[2]Department of Food Science and Technology, University of Reading, Whiteknights, Reading RG6 6AP, United Kingdom

Flavor extracts of four different varieties of fresh cherry fruit were obtained using a static vacuum extraction apparatus. The extracts were analysed using GC-MS and three chiral compounds were found in all the cherry varieties: limonene, linalool, and α-terpineol. Multidimensional GC (MDGC), incorporating a cyclodextrin based stationary phase, was used to separate the enantiomers of these chiral compounds. Enantiomeric excesses of *(R)-(+)*-limonene, *(R)-(–)*-linalool and *(R)-(+)*-α-terpineol were found. Each variety of cherry was canned and heat processed in a retort. The canned cherries were then extracted and analysed under similar conditions to the fresh samples. The canned cherries contained the same chiral compounds as found in the fresh. However, the concentration of the chiral compounds changed slightly on processing. Limonene showed a decrease in total concentration whereas α-terpineol increased. Moreover, analysis by MDGC showed a marked decrease in the enantiomeric excess of *(R)-(+)*-limonene, and small decreases for (R)-(–)-linalool and (R)-(+)-α-terpineol.

Chiral compounds are frequently found among the flavor volatiles of fruits and, like many naturally occurring chiral compounds, one enantiomer usually exists with a greater preponderance when compared with its antipode. Chiral odor compounds may show qualitative and quantitative differences in their odor properties (*1*). For example, (R)-(+)-limonene has an orange-like aroma while (S)-(–)-limonene is turpentine-like; (S)-(+)-carvone is characteristic of caraway while its enantiomer has a spearmint odor (*2*). However, other chiral compounds, such as γ- and δ-lactones, show very little enantioselectivity of odor perception (*1*). The occurrence of chiral flavor compounds in enantiomeric excess provides the analyst with a means of authenticating natural flavorings, essential oils, and other plant extracts. The advent of cyclodextrin-based gas chromatography stationary phases has resulted in considerable activity in the analysis of chiral compounds in flavor extracts of fruits, spices and other plants (*3-7*).

Processing of foods will cause qualitative and quantitative changes in the flavor volatiles. Some of these may give desirable flavor characteristics (*8*), while other changes may be undesirable (*9*). Relatively little is known about the effect of food processing on volatile flavor compounds in fruits and other foods, although many fruits are processed through canning or bottling, or in the production of jams and liqueurs.

The objective of this paper was to use a static vacuum simultaneous distillation-extraction system to obtain flavor extracts and examine any changes which may occur in the chiral compounds of cherries during processing (canning) using multidimensional gas chromatography.

**Materials and Methods**

Duplicate extractions were performed as follows on four different varieties of cherries which were obtained through a commercial importer. These were *Bing* (Chile), *Picota* (Spain), *Ruby* (USA), and *Garnet* (USA).

**Static Vacuum SDE.** Washed, de-stoned cherry flesh (100 g) was blended with 100 ml of deionised water. A 100 ml sample of this fruit slurry was transferred to the sample flask of a static vacuum SDE apparatus, as described by Maignial *et al* (*10*). Iso-octane (2.5 ml) was used as the extracting solvent. The temperature of the sample and solvent heating waters was 38°C and 22°C, respectively (±2°C). The cooling mixture was a 1:3 ethylene glycol:water mix and was maintained at -5°C (±1°C) by a chilling unit. After extraction, the solvent extract was concentrated to 500 µl using a rotary evaporator.

**Processing.** The cherry samples were canned in standard sized lacquered cans (7.5 cm diameter x 11 cm), filled with deionised water, and sealed. The cans were then heat treated at 121°C for 2.5 min (12 D concept) using a Fraser Vertical Retort with a Taylor MOD 30 Controller. The canned samples were then opened and volatiles extracted using the static vacuum SDE as described above.

**Gas Chromatography.** Extracts were analysed using a Carlo Erba 4160 Fractovap gas chromatograph equipped with a flame ionization detector. A Stabilwax-DA column (60 m x 0.32 mm ID; film thickness 0.5 µm; Restek Corp.) was used to separate the components of the extracts. The GC oven temperature was programmed at 30°C for 30 secs. and then heated at 40°C/min to 60°C, and from there at 3°C/min to 200°C. It was then held isothermally until the end of the run. The injector temperature was 250°C. Each extract (2 µl) was injected in the splitless mode and the injector chamber was purged after 30 sec. Quantitation of the chiral compounds was achieved using 1,2-dichlorobenzene as an internal standard. A 25 µl aliquot of a 1000 mg/kg 1,2-dichlorobenzene standard was added to the solvent before extraction.

**Gas Chromatography-Mass Spectrometry.** GC-MS analyses were carried out on a Hewlett-Packard 5890 Series II GC coupled to a VG7070E mass spectrometer (electron ionisation at 70eV; source at 200°C) with an Opus data handling system. A Stabilwax-DA column (60m x 0.32mm ID; film thickness 0.5µm ; Restek Corp.) was used to separate the components of the extracts. The GC temperature programme was similar to the Gas Chromatography conditions above.

**Multidimensional Gas Chromatography.** MDGC was performed on a linked dual oven system constructed in-house. The system comprised of a Carlo Erba 5160 Mega Series containing a pre-column, and a Carlo Erba Fractovap 4200 series GC containing the analytical column. Both GC's were fitted with flame ionisation detectors. A *MU*ltiple *S*witching *I*ntelligent *C*ontroller (*MUSIC*) heart cutting apparatus (Chrompack) was incorporated into the system. The *MUSIC* heart cutting device is based on the Deans pressure switching concept (*11*) and the switching manifold was located in the oven of the Carlo Erba 5160 Mega Series GC. The trapping device, cooled by liquid nitrogen, was located on Carlo Erba 5160 Mega Series GC. The pre-column was a Stabilwax-DA (30m x 0.53mm, film thickness 1.0μm; Restek Corp.) with a flow of 10 ml/min He, and the temperature of the FID was 250°C. The temperature programme used was the same as described in Gas Chromatography above. The analytical column was a Chirasil Dex CB column (permethyl-β-cyclodextrin, 25m x 0.25mm, film thickness 0.25μm, Chrompack) operating under a constant pressure of 0.88 bar, and the temperature of the FID was 240°C. The temperature program in the second GC differed for the three chiral compounds:

> limonene:        70°C for 15 min then at 4°C/min until elution
> linalool:        80°C for 15 min then at 4°C/min until elution
> α-terpineol:  120°C for 15 min then at 4°C/min until elution

Enantiomer configurational assignment was achieved by running enantiomerically pure standards. *(R)-(+)* and *(S)-(–)*-Limonene were obtained from Aldrich Chemical Co., Dorset, U.K.; *(R)-(–)* and *(S)-(+)*-linalool, *(R)-(+)* and *(S)-(–)*-α-terpineol were a gift from R.C.Treatt & Co. Ltd., Suffolk, U.K.

**Results and Discussion**

Gas chromatography (GC) has long been the method of choice for analysis for the flavor and fragrance industry due to the inherent volatility of the compounds of interest. Multidimensional GC (MDGC) is a modification of GC which involves trapping and further analysis of selected areas of the gas chromatogram (*12*). MDGC may incorporate a second GC oven containing an "analytical column" for further separation of compounds that have been trapped and transferred from the "pre-column" in the first GC. In recent years, MDGC has found particular application in the analysis of chiral compounds, where the "analytical column" is a capillary column coated with a cyclodextrin-based stationary phase (*13-15*).

Many foods contain components that are thermo-sensitive or labile and are changed by the application of heat resulting in artefact formation. One of the most commonly used extraction methods is simultaneous distillation-extraction (SDE). This method was first proposed by Likens and Nickerson in 1964 (*16*). Due to the high temperature used in atmospheric SDE, thermal artefacts may be generated during the extraction process and thus vacuum SDE systems have been developed to overcome the generation of thermal artefacts. One of the more recent vacuum SDE systems was developed by Maignial *et al* (*10*). This Static Vacuum SDE apparatus is an "easy-to-use device to isolate volatile components at room temperature, in a reasonable time, on a micro scale and with a high concentration factor". The SDE apparatus operates at pressure < 100 mbar, and therefore extraction is achieved at near room temperature. Thermal artefact generation was shown to be minimal and recoveries were similar to that found in atmospheric SDE.

After extraction of the fresh samples of cherries, 35 volatiles were identified by the GC-MS analysis of the flavor extract. The main components of the flavor extracts were hexanal, 3-methyl butanol, limonene, *trans*-2-hexenal, 1-hexanol, *cis*-3-hexen-1-ol, *trans*-3-hexen-1-ol, *trans*-2-hexen-1-ol, linalool, benzaldehyde, 1-octanol, benzyl acetate, benzyl alcohol, and α-terpineol. Due to the nature of the static vacuum extraction technique and the use of a high boiling solvent (iso-octane), compounds more volatile than hexanal were lost during concentration of the solvent extract in a rotary evaporator. This is one of the disadvantages of the static vacuum SDE technique and alternative methods for the analysis of highly volatile chiral compounds need to be used ( e.g., dynamic headspace followed by MDGC).

Benzaldehyde is considered to be the character-impact compound for cherry aroma (*17*), and is the key component in imitation cherry flavor compositions (*18*). The biosynthesis of this character-impact compound has been attributed to the acid or enzymic hydrolysis of the glucoside amygdalin in cherry pips. The importance of benzaldehyde in the aroma of cherry is emphasised by Schmid and Grosch (*19*). On sniffing stepwise diluted flavor extracts from cherries, seven compounds were revealed to have the highest aroma values - these compounds were benzaldehyde, linalool, hexanal, *trans*-2-hexenal, phenylacetaldehyde, (*Z,E*)-2,6-nonadienal and eugenol.

Three chiral compounds in the cherries (limonene, linalool and α-terpineol) were present at sufficiently high concentrations for further separation into their enantiomers using MDGC. Each component in turn was heart-cut onto the second "analytical" chiral column coated with permethyl-β-cyclodextrin where separation of the enantiomers was achieved. All chiral compounds showed an enantiomeric excess of the *R*- enantiomer (Table I). Small differences in the ratios of the enantiomer pairs between the four varieties were observed.

A marked change in the enantiomeric ratio of limonene was found on processing of the cherry samples (Table I). This was observed for all the varieties of cherries. Figure 1 shows MDGC chromatograms of the enantiomers of limonene from cherries before and after canning. Smaller changes in the enantiomeric ratios of linalool and α-terpineol were also observed. Racemization of chiral compounds during extraction and analysis of essential oils has been investigated previously (*20*). Linalyl acetate did not show any racemization during the extraction process, although it has been shown that linalool can racemise to some extent during atmospheric steam distillation extraction (*21*).

Limonene is known to be unstable in acid solution where it undergoes acid catalysed hydration to α-terpineol. It can also undergo thermal or UV induced oxidation to give p-cymene (Figure 2). A small decrease in the total concentration of limonene in the processed samples and a subsequent increase in the total concentration of α-terpineol shows that such reactions probably occurred during processing of the cherries.

Schmid and Grosch (*22*) have found an increase in the concentrations of benzaldehyde and linalool during jam preparation and also in simultaneous distillation-extraction of juices. According to these authors, the observed increase is attributed to hydrolysis of the β-glucoside and β-gentiobioside of hydroxyphenylacetonitrile present

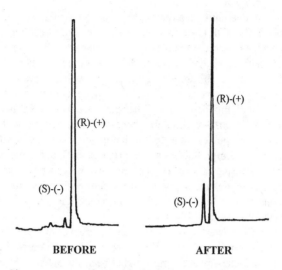

BEFORE          AFTER

**Figure 1.** Chromatograms from chiral MDGC analysis of limonene before and after processing (canning)

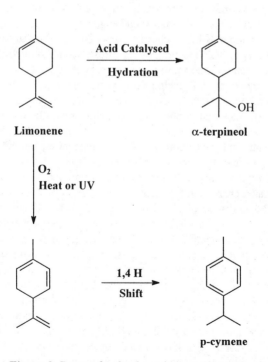

**Figure 2.** Routes for the degradation of limonene

Table I. Analysis of Chiral Compounds in Cherries before and after Processing

| Cherry Variety | | Limonene | | | Linalool | | | α-terpineol | | |
|---|---|---|---|---|---|---|---|---|---|---|
| | | (R) % | (S) % | Conc. mg/kg | (R) % | (S) % | Conc. mg/kg | (R) % | (S) % | Conc. mg/kg |
| Bing | Before | 93.4 | 6.6 | 41 | 74.9 | 25.1 | 21 | 62.5 | 37.5 | 20 |
| | | 92.2 | 7.8 | 35 | 74.7 | 25.3 | 19 | 62.3 | 37.7 | 21 |
| | After | 73.5 | 26.5 | 29 | 70.9 | 29.1 | 20 | 60.4 | 39.6 | 28 |
| | | 72.7 | 27.3 | 21 | 71.5 | 28.5 | 16 | 60.0 | 40.0 | 20 |
| Picota | Before | 88.6 | 11.4 | 41 | 80.5 | 19.5 | 28 | 71.4 | 28.6 | 29 |
| | | 88.2 | 11.8 | 49 | 79.7 | 20.3 | 28 | 70.8 | 29.2 | 28 |
| | After | 79.7 | 20.3 | 38 | 69.8 | 30.2 | 29 | 63.2 | 36.8 | 31 |
| | | 79.1 | 20.9 | 36 | 69.8 | 30.2 | 25 | 63.6 | 36.4 | 35 |
| Ruby | Before | 95.2 | 4.8 | 40 | 78.0 | 22.0 | 27 | 75.6 | 24.4 | 29 |
| | | 95.0 | 5.0 | 36 | 77.4 | 22.6 | 21 | 75.6 | 24.4 | 29 |
| | After | 79.1 | 20.9 | 33 | 70.6 | 29.4 | 24 | 68.1 | 31.9 | 31 |
| | | 78.3 | 27.1 | 33 | 70.4 | 29.6 | 26 | 68.3 | 31.7 | 29 |
| Garnet | Before | 92.1 | 7.9 | 48 | 81.0 | 19.0 | 28 | 78.7 | 21.3 | 37 |
| | | 92.1 | 7.9 | 50 | 81.2 | 18.8 | 32 | 78.1 | 21.9 | 33 |
| | After | 72.5 | 27.5 | 41 | 75.6 | 24.4 | 26 | 66.4 | 33.6 | 39 |
| | | 72.3 | 27.7 | 43 | 75.2 | 24.8 | 24 | 66.8 | 32.3 | 33 |

in the flesh of cherries for benzaldehyde, and to the hydrolysis of linalool glycoside for linalool (23-25).

It is unknown at present whether the decrease in enantiomeric excess in the processed cherry samples is due to partial racemization of the chiral compounds during processing catalysed by acid and/or the surface of the can, or whether chiral flavor compounds released from sugars during the processing conditions appear as either configurational stereoisomer. Further investigations are being carried out.

## Literature Cited

1. Ohloff, G., *Scent and Fragrances*, Springer-Verlag, Berlin, 1994.
2. Friedman, L.; Millar, J., *Science*, **1971**, *172*, 1044.
3. Ravid, U.; Putievsky, E.; Katzir, I.; Weinstein, V.; Ikan, R., *Flav. Frag. J.*, **1992**, *7*, 289.
4. Werkhoff, P.; Brennecke, S.; Bretschneider, W.; Güntert, M.; Hopp, R.; Surburg, H., *Z. Lebens. Unters.Forsch.*, **1993**, *196*, 307.
5. Kreis, P.; Mosandl, A., *Flav. Frag. J.*, **1993**, *8*, 161.
6. Mosandl, A.; Kustermann, A.; Guichard, E., *J. Chrom.*, **1990**, *498*, 396.
7. Bicchi, C.; Manzin, V.; D'Amato, A.; Rubiolo, P., *Flav. Frag. J.*, **1995**, *10*, 127.
8. Mottram, D. In *Thermally Generated Flavors*; Parliment, T.; Morello, M.; McGorrin, R., Eds.; ACS Symposium Series 543; ACS: Washington, D.C., 1992; 104-126.
9. Rouseff. R.; Nagy, S.; Naim, M.; Zehavi, U. In *Off-Flavours in Foods and Beverages*; Charalambous, G., Ed.; Developments in Food Science 28; Elsevier: Amsterdam, Netherlands, 1992; 211-227.
10. Maignial, L.; Pibarot, P.; Bonetti, G.; Chaintreau, A.; Marion, J., *J. Chrom.*, **1992**, *606*, 87.11. Deans, D., *Chromatographia*, **1968**, *1*, 18.
12. Schomburg, G.; Hausmanm, H.; Hübinger, E., *J. High Res. Chrom.*, **1984**, *7*, 404.
13. Schreier, P.; Bernreuther, A.; Christoph, N., *J. Chrom.*, **1989**, *481*, 363.
14. Bicchi, C.; D'Amato, A.; Frattini, C.; Nano, G.; Pisciotta, A., *J. High Res. Chrom.*, **1989**, *12*, 705.
15. Schomburg, G.; Weeke, F.; Müller, F.; Oreans, M., *Chromatographia*, **1982**, *16*, 87.
16. Likens, S.; Nickerson, G., *Proc. Am. Soc. Brew. Chem.*, **1964**, 5 - 13.
17. Schmid, W.; Grosch, W., *Z. Lebens. Unters. Forsch.*, **1986**, *183*, 39.
18. Broderick, J., *Int. Flav. Food Add.*, **1975**, *2*, 103.
19. Schmid, W.; Grosch, W., *Z. Lebens. Unters. Forsch.,* **1986**, *182*, 407.
20. Armstrong, D.; Li, W., *J. Chrom.*, **1990**, *509*, 303.
21. Hener, U.; Kreis, P.; Mosandl, A., *Chem. Mikrob. Technol. Lebensm.*, **1992**, *14*, 129.
22. Schmid, W.; Grosch, W., *Z. Lebens. Unters. Forsch.*, **1986**, *183*, 39.
23. Engel, K.; Tressel, R., *J. Agric. Food Chem.*, **1983**, *31*, 958.
24. Heidlas, J.; Lehr, M.; Idstein, H.; Schreier, P., *J. Agric. Food Chem.*, **1984**, *32*, 1020.
25. Salles, C.; Essaied, H.; Chalier, P.; Jallageas, J.; Crouzet, J. In *Bioflavour '87*; Schreier, P., Ed.; W. de Grutyer: Berlin, 1988; 145 - 160.

# Chapter 8

# Glucosone as a Radical-Generating Intermediate in the Advanced Maillard Reaction

**S. Kawakishi[1], S. Nasu, R. Z. Cheng, and T. Osawa**

**Department of Applied Biological Sciences, Nagoya University, Nagoya 464−01, Japan**

Amadori compound, a most important intermediate in Maillard reaction, is transformed to several final products, melanoidin, flavors and some fluorescent product (AGE) through α-dicarbonyl compounds. One of them, D-glucosone (D-*arabino*hexosulose), was also formed by the oxidative degradation of Amadori compound under the presence of trace amounts of transition metal ion and oxygen. In this oxidation process, some active oxygens, superoxide, hydrogen peroxide and hydroxyl radical, were generated and these species participated to the oxidation of food and bio-materials. Glucosone arose the oxidative fragmentation and polymerization of protein under the presence of transition metal ion. Moreover, glucosone-metal ion system also oxidized histidine and tyrosine residues in protein molecule. These oxidative damages of amino acid and protein would be induced with oxygen radicals generated through glucosone-metal ion-oxygen complex.

It is well known that Maillard reactions between carbonyl and amino groups are usually arisen anywhere reducing sugar and amino acid or protein occur like foods and biological tissues. Most important product in an initial stage of Maillard reaction is Amadori compound which is generated through proton rearrangement of Schiff's base from glucose and amino acid. In food processing, the oxidative degradation of Amadori compound induces the formation of melanoidin and many volatile compounds. Many intermediates in the process to some endproducts from Amadori compounds are identified (*1*) and among them, α-dicarbonyl compounds, such as glucosone, 3-deoxyglucosone(3-DG) and 1-deoxyglucosone(1-DG), are known as most active species for the advanced reactions(Figure 1). We have found the preferential formation of glucosone from Amadori compounds under the presence of samll amounts of transitional metal ions(*2,3*). Moreover, glucosone is quickly degraded to lower molecullar sugar components by the action of metal ions and generated some oxygen radical species at the same time(*4,5*). These active oxygens contain a series of products formed with one electron reduction of oxygen molecule, that is, superoxide, hydrogen peroxide and hydroxyl radical(*6*). In other words, the

[1]Current address: Yawata 70, Miyoshi-cho, Nishikamo-gun, Aichi-ken 470−02, Japan

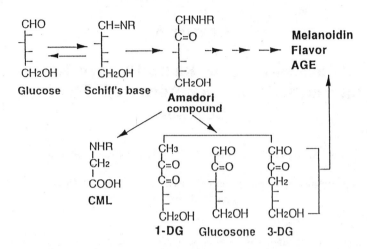

**Figure 1.** Initial stages of Maillard reaction.    R=amino acid or protein.

Amadori compound or glucosone-transition metal ion under the presence of oxygen are the active oxxygen generating systems. Therefore, when these systems coexist with other food components as polysaccharides and proteins, these conditions are commonly in foods, these food components would be suffered some oxidative damages containing their degradation and polymerization through radical reaction.

## Formation of Glucosone from Amadori Compound with Copper Ion

As an Amadori compound, Nα-t-Boc-Nε-fructoselysine(fructoselysine, FL), prepared from Nα-t-Boc-lysine and glucose by the method of Njoroge et al(7) were used. FL(10mM) was incubated with cupric ion(50μM) as transition metal ion in phosphate buffer solution (1/15M, pH 7.2) at 40℃ and then, about 70% of FL was gradually degraded to lysine and aldosulose during 24h(2,3). Their yields were approximately equimolar. Moreover, most of aldosulose was glucosone, and 1-deoxyglucosone(1-DG) and 3-deoxyglucosone(3-DG) were minor. However, the same oxidative degradation of FL has been observed at long time incubation in buffer solutions without metal ion.    FL(10mM) was incubated with    phosphate(0.2M, pH 7.4), Tris-HCl buffer(0.2M, pH 7.4) and aqueous solution at 45℃ for 24 days without addition of cupric ion. The decomposition of FL and the formation of lysine and glucosone were arisen more slowly than the case of presence of cupric ion. Especially, these reactions were most remarkable in phosphate buffer solution and they were markedly inhibited by the addition of DTPA(0.1mM). Therefore, the oxidative degradation of FL would be derived autoxidatively by trace amount of metal ion presented in buffer solution. From these results, Amadori compound must be degraded through the following process : Amadori compound may forms a complex of Amadori compound-metal ion-oxygen, since it have a strong affinity with transition metal ion, and the electron transfer would originate from Amadori compound to oxygen via metal ion, and C-N bond of Amadori compound is oxidatively cleaved to give amino acid and aldosulose.

Since these autoxidative degradation of Amadori compound with participation of metal ion necessarily generates some active oxygen species, some oxidative damages would be arisen in food and bio-materials coexisted with Amadori compound. We have already published the papers that protein have been oxidatively degraded to lower molecular peptides with Amadori compound-cupric ion system(8).

## Oxidative Damage of Protein and Amino Acid Induced by Glucosone-Cupric Ion System

$\alpha$-Dicarbonyls as glucosone equilibrates partially with their enediol structure and the enediol form of glucosone may easily bind with transition metal ion similar to ascorbic acid and the enaminol form of Amadori compound to give oxygen radical through electron transfer from glucosone to oxygen. If this hypothesis is correct, the glucosone-cupric ion system may induce some oxidative damage of protein. Then, the mixture of glucosone(5mM), cupric ion(50μM) and bovine serum albumin (BSA,0.04%) in phosphate buffer solution(1/15M, pH 7.2) was incubated at 37℃ for 24h and the time dependent changes of BSA were investigated by SDS-PAGE. BSA was fragmented to lower peptides within 4h and this fragmentation was markedly inhibited by the addition of EDTA and catalase(9). In both cases of BSA-cupric ion and BSA-glucosone systems as controls, any changes in BSA were not observed within 24h of incubation(8,9).

These results suggested that glucosone-cupric ion complex generated hydrogen peroxide from oxygen by its reduction through superoxide and these active oxygens or some oxygen-cupric ion complex attacked to protein molecule to arise its protein fragmentation.

## Oxidative Degradation of Histidine Residue in Protein and Peptide with Glucosone and Cupric Ion.

On protein fragmentation in BSA with glucosone-cupric ion system, the remarkable decrease of histidine residue in BSA was observed by amino acid analysis of the oxidized BSA (over 80% of native amount on their incubation for 24h). Then, the detail investigation on histidine degradation was performed by using of histidine derivative and histidine containing peptides. The reaction mixtures of peptides(1mM) or N-benzoylhistidine (1mM), glucosone(5mM) and cupric ion(50μM) in phosphate buffer solution (1/15M, pH7.2) were incubated at 37℃ for 24h and the time-dependent changes of peptides and histidine derivative were determined by HPLC method. As shown in Fig. 2, about 50% of β-alanylhistidine and histidyltyrosine were degraded during 24h and especially, the decrease of N-benzoylhistidine was most considerable than two peptides. The time-dependent changes in amino acids were investigated by amino acid analyses. These results showed in Table I and II. The formation of ammonia and aspartic acid with the decomposition of histidine   suggests the oxidative cleavage of its imidazole ring. On the other hand, histidine residues did not change during the incubation in absence of cupric ion. N-Benzoylhistidine   treated with ascorbate-cupric ion system was markedly oxidized to 2-oxohistidine (peak 3 in Fig. 3 a) which was degraded to final product, aspartic acid,  through the cleavage of its imidazole ring *(10,11)*. On the condition replaced Cu(II) with Fe(II), the same results have been obtained*(12)*.

In addition, for establishing the oxidative degradation of histidine in glucosone mediated system, the oxidized sample obtained from Table I was analyzed by HPLC of same condition used the case of ascorbic acid-cupric ion. As shown in Fig. 3 b, HPLC profile was perfectly agreed with that of ascorbate mediated system. Fig. 4 showed the chemical structure of characterized products in Fig. 3 and the proposed mechanism for the ring cleavage of imidazole ring *(9)*. Moreover, the same

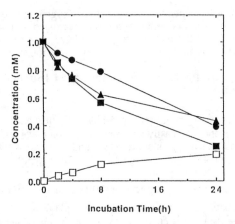

**Figure 2.** Time-dependent degradation of histidine and histidine-containing peptides. ● β-alanylhistidine, ▲ histidyltyrosine, ■ N-benzoylhistidine, □ 2-Oxo-N-benzoylhistidine.
(Reproduced with permission from reference 9. Copyright 1992 The Biochemical Society and Portland Press.)

**Table I Time dependent Changes in Amino Acid Composition of the Oxidized N-Benzoylhistidine with Glucosone-Cu(II) System**

| Amino Acid | Molar Ratio (%) | | | | |
|---|---|---|---|---|---|
| | 0 | 2h | 4h | 8h | 24 h |
| Asp | 0 | 2.0 | 2.9 | 5.6 | 11.5 |
| Gly | 0 | 0.7 | 0.7 | 2.3 | 1.6 |
| His | 70.2 | 67.4 | 59.4 | 41.1 | 24.0 |
| NH$_4$OH | 29.8 | 27.8 | 37.7 | 41.1 | 63.0 |

(Reproduced with permission from reference 9. Copyright 1992 The Biochemical Society and Portland Press.)

**Table II Time Dependent Changes in Amino Acid Composition of the Oxidized Histidyltyrosine with Glucosone-Cu(II) System**

| Amino Acid | Molar Ratio (%) | | | | |
|---|---|---|---|---|---|
| | 0 | 2h | 4h | 8h | 24h |
| Asp | 0 | 0.2 | 0.3 | 0.3 | 0.5 |
| Glu | 0 | 0.3 | 0.4 | 0.5 | 0.8 |
| Gly | 0 | 0.4 | 0.5 | 0.7 | 1.2 |
| Tyr | 47.0 | 46.9 | 47.6 | 48.8 | 49.1 |
| His | 46.8 | 43.8 | 42.3 | 37.4 | 32.2 |
| NH$_4$OH | 6.1 | 9.3 | 9.8 | 13.7 | 18.7 |

(Reproduced with permission from reference 9. Copyright 1992 The Biochemical Society and Portland Press.)

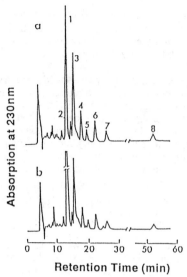

**Figure 3.** HPLC of the Oxidation Products of N-Benzoylhistidine with Ascorbate-Cu(II) (a) and Glucosone-Cu(II) (b) Systems,
HPLC Condition; Develosil ODS-5 (8x250mm); 0.1%TFA/MeOH=3/1(v/v); Flow Rate, 1.5mL/min; Detection at 254nm.
(Reproduced with permission from reference 9. Copyright 1992 The Biochemical Society and Portland Press.)

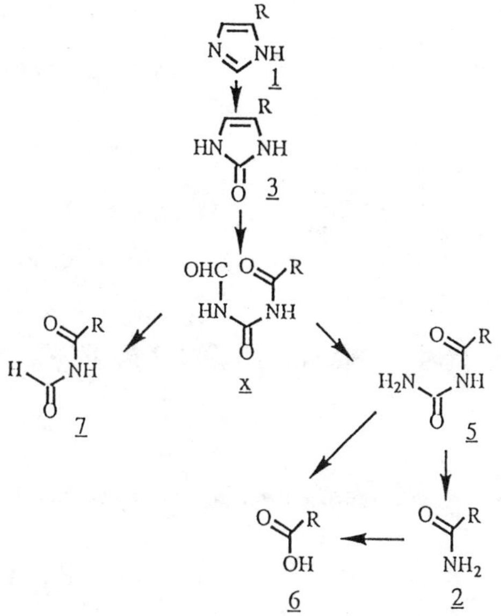

**Figure 4.** Proposed Mechanism for the Oxidative Cleavage of Imidazole Ring in Histidine with Glucosone-Cu(II) System.
(Reproduced with permission from reference 9. Copyright 1992 The Biochemical Society and Portland Press.)

degradation of histidine residue in protein has been observed on the reaction mediated with $H_2O_2$-Cu(II) *(13)*. Therefore, glucosone generates hydroxyl radical or active copper-oxygen complex through the formation of glucosone-Cu(II)-$O_2$ complex, and such active species may attack to imidazole ring of histidine.

**Formation of Dityrosine in Protein by the Oxidative Action of Glucosone and Cupric Ion.**    As the action of hydroxyl radical and metal catalyzed systems to protein, its polymerization or aggregation is well known besides its fragmentation *(14,15,16)*. Some kind of crosslinker must be produced by radical reaction for intermolecular linking of protein. On our preliminary experiments, lysozyme easily polymerized by the action of several oxygen radical-generating systems. Then, we have investigated on the polymerization of lysozyme and its crosslinker with FL and glucosone-Cu(II) systems. The reacion mixture composed of lysozyme(0.04%), FL(5mM) or glucosone (5mM) with and without cupric ion(50μM) in phosphate buffer solution(1/15M, pH 7.2) was incubated at 37℃ for 24h. The changes of lysozyme were checked in time course by using of SDS-PAGE. From their results shown in Fig. 5, lysozyme dimer was predominantly formed after 2h and in addition to dimer, minor level of trimer and polymer were observed with incubation time. Amino acid analyses of the oxidized lysozyme exhibited the considerable decreases of methionine and tyrosine residues. The decrease of tyrosine was partially due to the formation of dityrosine by HPLC of hydrolyzate of the oxidized lysozyme. Dityrosine formation in lysozyme oxidized by FL-Cu(II) system was shown in Fig. 6 but dityrosine was not formed without cupric ion.

　　　These oxidation was perfectly suppressed by the addition of EDTA(0.1mM), catalase(500units/mL) and aminoguanidine(10mM) and so it suggests that the dimerization of tyrosine evidently proceeds with radical reaction. Dityrosine has been

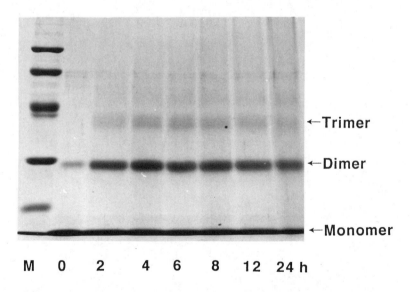

**Figure 5.**    Formation of Lysozyme Polymer with Glucosone-Cu(II) System.
M : Marker, (from lower to upper) 14, 20, 30, 43, 67, 94 Kda.

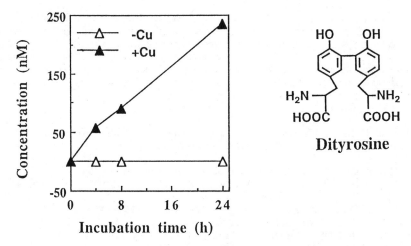

**Figure 6.** Formation of dityrosine in oxidized lysozyme with glucosone-Cu(II) system. Dityrosine determination by HPLC; Develosil ODS-5 (4.6x250mm) ; 0.5%AcOH/MeOH (29/1(v/v); flow rate 0.8mL/min; Detection, fluorescence, Ex 283nm, Em 403nm.

confirmed as a typical marker of the oxidative damage in protein with radiolytic and metal-catalyzed oxidation*(17,18)*.

## Degradation of Glucosone with Cupric Ion and Oxygen

As mentioned above, glucosone-Cu(II)-$O_2$ system generates active oxygen species which relate to  the oxidative damage of food and bio-materials. In this case, the oxidative degradation of glucosone to lower saccharide must be commonly arisen. Glucosone(5mM)-Cu(II) was incubated in phosphate buffer solution(1/15M, pH 7.2) at 37°C and the samples taken after 1 and 7 days were reduced, acetylized and followed by GC-MS analyses. GC analyses showed that glucosone was degraded to tetrose and triose through pentosulose (data not shown). However, when DTPA was added to above reaction mixture, the degradation of glucosone stayed on pentosulose, but not to tetrose and triose. Each step to triose from glucosone under the presence of transition metal ion and oxygen was also the processes on spontaneous generation of active oxygen species.

## Conclusion

Amadori compounds formed on the initial stage of Maillard reaction are  transformed to melanoidin and flavor in food processing and sometime to fluorescent products(AGE) in biological tissues. As intermediates on the degradation process of Amadori compounds, several α-dicarbonyl compounds are identified, and among them, glucosone has been selectively produced from Amadori compounds under the presence of trace amounts of transition metal ion. Moreover, the formation and degradation process of glucosone with metal ion may accompany the generation of active oxygen species, superoxide, hydrogen peroxide, hydroxyl radical and etc. These facts have been indirectly demonstrated by the experiments on the oxidative damages of protein and amino acids with glucosone-Cu(II)-$O_2$ system and the scavenger effects for these reactions.

These oxidation reactions usually relate on food deterioration in processing. However, there are several antioxidative vitamin in food, ascorbic acid, tocopherol and carotene which attack to active oxygens as radical scavenger(19). Besides these antioxidative vitamin, many phenolic compounds, namely flavone, catechin, anthocyanidine, tannin and etc, scavenge or quench active oxygen species and trap metal ion to protect food nutrient materials from their oxidative damage(20,21).

## Literature Cited

1. Ledl, F. In *The Maillard Reaction in Food Processing, Human Nutrition and Physiology*; Fino, P.A., et al Eds.; Birkhäuser Verlag; Basel, 1990; 19-42.
2. Kawakishi, S.; Tsunehiro, J.; Uchida, K. *Carbohyd. Res.* **1991**, *211*, 167.
3. Cheng, R.Z.; Tsunehiro, J.; Uchida, K.; Kawakishi, S. *Agric. Biol. Chem.* **1991**, *55*, 1993.
4. Cheng, R.Z.; Kawakishi, S. *J. Agric. Food Chem.* **1993**, *41*, 361.
5. Cheng, R.Z.; Kawakishi, S. *J. Agric. Food Chem.* **1994**, *42*, 700.
6. Fridovich, I. In *Free Radicals in Biology*; Pryor, W.A. Ed.; Academic Press; New York, 1976, *1*, 239-277.
7. Njoroge, F.G.; Fernandes, A.A.; Monnier, V.M. *J. Biol. Chem.* **1988**, *263*, 10646.
8. Kawakishi, S.; Okawa, Y.; Uchida, K. *J. Agric. Food Chem.* **1990**, *38*, 13.
9. Cheng, R.Z.; Uchida, K.; Kawakishi, S. *Biochem. J.*. **1992**, *285*, 667.
10. Uchida, K.; Kawakishi, S. *Agric. Biol. Chem.* **1988**, *52*, 1529.
11. Uchida, K.; Kawakishi, S. *Bio-org. Chem.* **1989**, *17*, 330.
12. Levine, R.L. *J. Biol. Chem.* **1983**, *258*, 11823, 11828.
13. Uchida, K.; Kawakishi, S. *J. Agric. Food Chem.* **1990**, *38*, 660.
14. Davis, K.J.A. *J. Biol. Chem.* 1987, *262*, 9895.
15. Kano, Y.; Sakano, Y.; Fujimoto, D. *J. Biochem.* **1987**, *102*, 839.
16. Dermott, M.J.; Chiesa, R.; Spector, A. *Biochem. Biophys. Res. Commun.* **1988**, *157*, 626.
17. Giulivi, C.; Davis, K.J.A. *J. Biol. Chem.* **1993**, 268, 8752.
18. Huggins, T.G.; Well-Knecht, M.C.; Detorie, N.A.; Baynes, J.W.; Thorpe,S.R. *J. Biol. Chem.* **1993**, *268*, 12341.
19. Niki, E. In *Vitamin E;* Mino, M., et al Eds.; Karger; Basel, 1992; 23-30.
20. Okuda, T.; Yoshida, T.; Hatano, T. In *Active Oxygens, Lipid Peroxides, and Antioxidants;* Yagi, K., Ed.; Japan Scientific Societies Press, 1993; 336-346.
21. Namiki, M.; Yamashita, K.; Osawa, T. In *Active Oxygens, Lipid Peroxides, and Antioxidants*; Yagi, K., Ed.; Japan Scientific Societies Press,1993; 319-332.

# CHEMICAL MARKERS
# OF STORAGE EFFECTS

# Chapter 9

# Chemical Degradative Indicators To Monitor the Quality of Processed and Stored Citrus Products

H. S. Lee and S. Nagy

Citrus Research and Education Center, Florida Department of Citrus, 700 Experiment Station Road, Lake Alfred, FL 33850

This overview deals with some of the chemical marker compounds suited to the requirements of quality control laboratories in the citrus industry. Chemical markers include    ascorbic acid, dehydroascorbic acid, hydroxymethylfurfural, furfural, 2,5-dimethyl-4-hydroxy-3(2H)-furanone, 2,3-dihydro-3,5-dihydroxy-6-methyl-4H-pyran-4-one, 4-vinyl guaiacol and $\alpha$-terpineol. Some of these compounds might be applied as useful tools in evaluating quality deterioration due to unsuitable manufacturing and storage as well as for the computation and optimization of  the manufacturing processes and parameters.  Since they are all chemically reactive compounds, careful evolution of  kinetics of these compounds is necessary.

Ascorbic acid degradation, nonenzymic browning, acid-catalyzed hydration-dehydration and other chemical  reactions are generally believed to cause a loss of freshness, discoloration and  formation of off-flavors in citrus. From extensive studies with stored citrus juice, and model systems simulating citrus juices, many  compounds are considered   primarily  responsible  for the quality deterioration of  citrus juices. Some of them are considered as chemical degradative indicators that might provide the basis for a definitive test of citrus  juice quality  because their analytical methods are well established and their baseline concentrations would be either zero or very low in citrus juices. It is our intention to discuss  chemical degradative indicators that have been tested on citrus juices, and state what we believe to be their weaknesses and strengths.

## Ascorbic Acid and Its Degradation Products

Ascorbic acid (AA) is one of the principal nutritional components of citrus juices, and

0097–6156/96/0631–0086$15.25/0

many consumers purchase and consume citrus juices because of this vitamin. Ascorbic acid stability, its influence on quality during storage, and its influence on processing citrus juices were subjects of active study. Ascorbic acid is highly sensitive to heat, alkali, oxygen, and light. Consequently, appreciable amounts of ascorbic acid can be destroyed by juice processing, packaging and subsequent storage due to oxidative and thermal reactions. Thus, quantitative evaluation of ascorbic acid content was considered as one of the simple approaches to evaluate processing and storage parameters, and to predict the shelf-life of citrus products. Ascorbic acid degradation in citrus juices is also believed to be linked to nonenzymic browning. In the nonenzymic browning reactions, ascorbic acid and its degradation products function as reactive carbonyl components.

**L-Ascorbic Acid.** The degradation of ascorbic acid is known to occur by both oxidative and nonoxidative mechanisms and is generally characterized as a first-order reaction. The rate of nonoxidative loss of AA is often one-tenth (*1*) or up to one-thousandth (*2*) the rate of loss under aerobic conditions . Smoot and Nagy (*3*) gave extensive data on ascorbic acid loss for grapefruit juice and orange juice (*4*) stored at various temperatures in hermetic containers. Nagy and Smoot (*4*) have shown that storage temperatures in excess of 28°C caused vitamin C destruction at markedly accelerated rates in canned products. The activation energy (Ea) for ascorbic acid degradation in single-strength grapefruit juice (SSGJ) was 18.2 Kcal/mol and $Q_{10}$ of the reaction was about 2.7 (*3*). In contrast to grapefruit juice, Ea was 24.5 Kcal and $Q_{10}$ was 3.7 in single-strength orange juice (SSOJ) (*4*). A recent study (*5*) for the kinetics of ascorbic acid loss in orange juice as functions of concentration and temperature fitted to first-order kinetic model and activation energies for degradation in juice were on the order of 27.6 Kcal/mol, which is close to the value reported by Nagy and Smoot (*4*). Those kinetics of rate deterioration may be applicable to predict shelf-life of citrus juices in various conditions, but ascorbic acid loss in orange juice was known to be greater than in grapefruit juice at similar storage temperatures.

Figure 1 shows some quality parameters for storage-aged orange juice concentrate. Since ascorbic acid is quite unstable, its loss is sometimes used as an index of loss of overall quality in citrus juices. Figure 1a shows the gradual decrease of AA during storage of clarified, concentrated orange juice (serum) at 24°C and 5°C. At the beginning of storage, ascorbic acid content in the serum was about 32 mg/100mL and after 22 wks of storage, we found less than 16 mg/100mL in the serum. Thus, at the storage of 24°C, more than 50% of ascorbic acid was lost after 22 weeks. But changes in ascorbic acid content in the serum stored at 5°C were not significant compared to serum stored at 24°C. It retained more than 91% over the 22 weeks of storage. Thus, changes in ascorbic acid was noticeable at ambient temperature, indicating that high concentrated orange serum was not stable at ambient temperature for extended storage. But the changes in serum stored at 5°C were minimal, which might have economic feasibility for replacement of conventional storage of orange juice concentrates at freezing temperature (*6*).

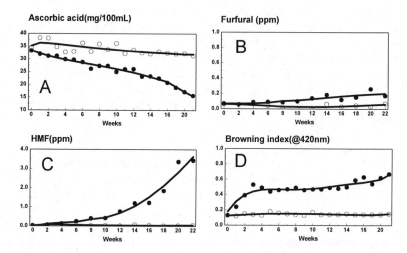

Figure 1. Some quality parameters for clarified orange juice concentrate stored at 25°C (●) & 5°C (O). (A) ascorbic acid; (B) furfural; (C) HMF; (D) browning index.

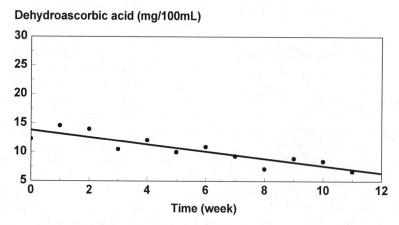

Figure 2. Changes of dehydroascorbic acid in EVOH-barrier pack of orange juice stored at 4.4°C.

**L-Dehydroascorbic Acid.** The first chemically stable product in the oxidation of L-ascorbic acid is L-dehydroascorbic acid (DHAA). A significant relationship between DHAA and browning of citrus juices was suggested by Kurata et al. (*7*) and Sawamura et al. (*8, 9*). Sawamura et al. (*8*) and Li et al. (*10*) indicated that DHAA, as a substantial starting material in citrus browning, degrades to form brown pigment even if oxygen is present or not in the juice. Thus, it may be important to prevent the degradation of DHAA in juice processing. In the previous storage test with grapefruit juices (*3*), the concentration of DHAA was found to have remained essentially unchanged during storage. In a recent study with orange juice, Hoare et al. (*11*) observed that DHAA levels increased significantly over the shelf life. After approximately 35 days at 3°C, nearly all of the vitamin C in the opened cartons of orange juice was in the form of DHAA. Figure 2 shows the result from the recent test with orange juice packed in EVOH-barrier package and stored over 12 weeks at 4.4°C. Gradual changes of DHAA were observed (Figure 2). Under the same storage conditions of temperature and water activity ($a_w$), it has been observed there are no significant differences in the rate constants for total AA (AA+DHAA) and AA destruction (*12*).

Even though DHAA has the same vitamin C activity as AA and its importance in citrus browning has been proposed in many previous studies, DHAA has not been widely adopted as a quality indicator when compared to AA. It may be due to the fact that DHAA is not stable and spontaneously converts to 2,3-diketo-L-gulonic acid (DKG) which has no antiscorbutic potency, and also its concentration is known not to be a significant fraction of ascorbic acid throughout the storage period. In citrus juices, both oxidative and nonoxidative pathways are operative during storage, and DHAA is the first oxidative degradation product but not under anaerobic condition. Thus, since large quantities of DHAA are present in citrus juice, it can be speculated that the oxidative pathway must be dominant.

Since most shelf-life studies of citrus juice products are concerned with the effects of processing and storage on vitamin retention, in particular vitamin C, many chromatographic methods are available for vitamin C measurement as ascorbic acid content. Ascorbic acid has absorption maxima at 245nm (log $\epsilon$ = 3.88) in acidic solution (*13*) and can be easily determined by 2,6-dichloroindophenol dye titration method or by reversed phase HPLC or by anion exchange mode HPLC. Dilution of orange juice with metaphosphoric acid (2-5% w/v) is a very common approach for extraction, and it also affords excellent stability during analysis. Dehydroascorbic acid has no defined absorption spectra in the near UV (*14*), and is electrochemically inactive; thus, it is known to be difficult to measure dehydroascorbic acid contents in citrus products. A complicated procedure for DHAA in citrus juices is based on the production of derivatives of DHAA with either dinitrophenyl hydrazine (DNP) or O-phenylenediamines (OPDA), or DHAA can be measured indirectly after its reduction to AA with reducing reagents such as D,L- homocysteine (*11*), D,L-dithiothreitol (*15*), or after the native AA has been oxidized to DHAA.

## Furans

Many furan derivatives have been identified as degradation products associated with browning in citrus products and model systems. Generally, furan derivatives are considered important aroma constituents from a sensory point of view. They are mainly associated with sweet, fruity, nutty or caramel-like odor impressions. Furans apparently determine the overall browning of storage-abused citrus products (16, 17). Among them, hydroxymethylfurfural (HMF) and furfural are two major furan derivatives in citrus juices, and both HMF and furfural are considered as useful indicators of temperature abuse in citrus juices.

**Furfural.** Furfural is the most widely used to monitor citrus juice quality because the furfural content of freshly processed juice is virtually zero whereas large amounts accumulate in storage-abused juice (18). Furfural in citrus juices seems to be derived from the decomposition of ascorbic acid; anaerobic degradation of ascorbic acid is thought to be responsible for approximate quantitative yields of furfural in citrus juices. Also, even though furfural does not contribute to the malodorous property of aged citrus juice, it is useful as an off-flavor indicator for temperature abuse of citrus juice (18,19). Especially for citrus juices with aseptic packaging, the use of furfural content as a flavor indicator becomes more important to food processors as aseptic processing and packaging of citrus juices becomes accepted (20). The reported aroma threshold value for furfural in orange juice is about 80 ppm (21) which is considerably higher than the level reported from aged citrus juices. In our storage test, it could not be reached even after 15 weeks at 50 °C storage in bottled grapefruit juice (17). Thus, it has limited use in correlating flavor change in stored citrus juices because of its high aroma threshold value.

Furfural has an especially significant relationship to browning (16), but careful observation was suggested for the estimation of rate of furfural accumulation to predict storage temperature abuse in orange juice concentrate. Figure 1b shows the gradual increase of furfural in juice concentrates as browning increases. Furfural increased more than 6 times, from 30 ppb to 190 ppb after 22 weeks storage at 24°C, while at 5°C, about 70 ppb of furfural was detected. Furfural accumulation appeared to be affected by storage temperature, time and deaeration (22) and 12°C seems to be a critical storage temperature, below which furfural accumulation is very slow (23). The kinetics of furfural formation in orange juice described as zero-order reaction, after a short induction period (24). Its mode of formation is described as specific acid catalysis reaction beacuse the reaction velocity is directly proportional to the hydronium ion activity (24).

The effect of juice concentration on furfural accumulation is noted in Figure 3; the accumulation of furfural is higher than its decomposition in single-strength juice; thus, furfural can serve as a quality deterioration index in this product. However, in concentrates furfural forms and decomposes simultaneously. Therefore it would be more difficult to use it as an index for predicting product quality changes in orange juice concentrate (22). From a previous study of furfural accumulation in both aseptically filled juice and concentrates, Kanner et al. (22) indicated the rapid decomposition of furfural in orange juice concentrates as the juice concentration increased. Furfural is a very reactive aldehyde; it can react with many other compounds in juice. Thus, the formation of furfural in citrus juices appears to be dependent on temperature, time, pH, and concentration of juice as well as the vitamin C content.

**Hydroxymethylfurfural.** Hydroxymethylfurfural, or HMF is a cyclic aldehyde formed by dehydration when hexoses are heated in acid solution, and is recognized as a primary breakdown product during the dehydration of glucose or fructose in an acid medium (25). HMF content is practically zero in fresh, untreated juice (26); thus, HMF content is suggested as a quality criterion (27) for a wide range of products including fruit juices, tomato products, soft drinks and honey. Even in properly manufactured juice concentrates, HMF may be formed by acid-catalyzed sugar dehydration, but high concentration of HMF in juice is considered to be due to excessive heat treatment during concentration or pasteurization.

In citrus products, the detection of HMF is suggested as the first indication of storage change even though it is not responsible for the off-flavor considered characteristic of stored citrus juice (28). Furthermore, HMF is known as a precursor of browning and its close relationship to the color deterioration in stored grapefruit juice has been determined (16). In fruit juice, concentration and type of sugars, quantity and structure of amino acids, pH, as well as packaging, storage and reaction with other compounds could also affect HMF formation during processing and subsequent storage of fruit juices (29). In our sugar-catalyzed model systems (30) simulating grapefruit juice, fructose was the major potential source for the formation of HMF. The rate of formation of HMF was observed as a function of temperature and soluble solids content (31). Its mode of formation was described as a second-order autocatalytic mechanism (32), by zero-order kinetics (33), and by zero-order with an induction period (34). In our previous test with single-strength canned and bottled grapefruit juices (16, 17), considerable amounts of HMF were formed as a function of temperature and length of storage periods. From the recent study of kinetics and mechanism of HMF formation (35), Amadori rearrangement product (ARP) is known to provide a low-energy pathway for the formation of HMF compared to the pathway from glucose alone. Activation energy (Ea, Kcal mol$^{-1}$ K$^{-1}$) for the HMF formation from ARP was 4.10-4.15 compared to 25.8 from glucose.

Figure 1c shows the gradual build-up of HMF contents in clarified 80°Brix orange juice concentrate at 5 and 24 °C. There was about 0.03 ppm level of HMF from the initial sample and we found more than 3 ppm HMF after 22 weeks storage at 24°C. At the storage temperature of 5°C, less than 0.2 ppm was found in concentrate after 22 weeks of storage. Thus, concentrate stored at 24°C contained about 15 times more HMF than the concentrate stored at 5°C. In the relationship between HMF content and sensory qualities of juice, if values of HMF of about 20 ppm and more occur together with a cooked or bready flavor in grape juice and apple juices (36), excessive thermal stress is indicated. In the apple juice industry, 30 ppm of HMF is used as a reasonable limit for commercial acceptability (31). However, this limit seems unreasonably high for citrus juices because this critical region is only reached after 3 weeks at 50°C; under these storage conditions, significant color changes compared to control would have already occurred (16) in stored grapefruit juice. In citrus juices,

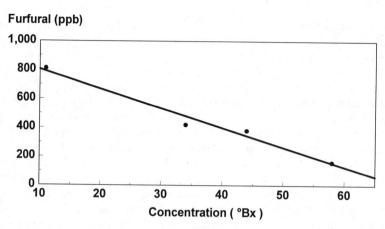

Figure 3. Effects of concentration on furfural (ppb) formation and degradation in orange juice. Adapted from Ref. 22.

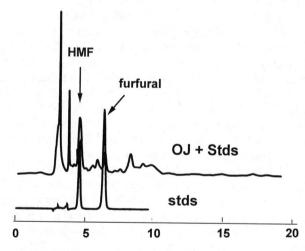

Figure 4. HPLC chromatogram of HMF and furfural in orange juice.

HMF has been correlated well with browning in orange juice (*17, 37*), but no correlation was found between HMF and sensory quality deterioration of citrus juices (*14*). From the previous storage test with grapefruit juices over a 15 week period at temperatures ranging from 10 to 50°C, only at temperatures of 40 and 50°C did HMF exceed its flavor threshold of 100 ppm (*16*). Data in the previous study of aseptically packed orange juice concentrate (60°Bx) during storage at 25°C (*38*) showed that browning and taste score were more consistent indicators for quality changes than HMF and ascorbic acid degradation. Thus, HMF has been of only limited use in correlating the occurrence of atypical smell and taste in stored citrus juice.

Several methods including spectrophotometric or chromatographic techniques have been proposed for the determination of furans such as HMF or furfural, or developed to determine both furan compounds simultaneously. The spectrophotometric methods do not differentiate between HMF and furfural without the need of a previous separation procedure. The commonly employed colorimetric method for HMF is Winkler's (*38*), which is based on the use of barbituric acid and p-toluidine. Toxicity of p-toluidine, instability of the color complex formed, and interference of sulfurous acid and possibly other compounds present in the fruit juice (*40*) are known to be problems. Unlike HMF, the colorimetric method for furfural was based on a previous distillation of juice and the colorimetric analysis of distillate for furfural based on colorimetric reaction with aniline acetic acid in juices (*41*). However, distillation procedure with poor recovery (about 34%) and long reaction time for color development (approximately 1 h), and requirement of hazardous chemical aniline (*20*) are known to be the drawbacks of the colorimetric method for furfural. Also, colorimetric method generally requires strict control of reaction time and temperature to achieve stable and reproducible color development.

HPLC procedures (*16, 20, 43, 44, 45*) offer more precision along with high selectivity and sensitivity as compared to the colorimetric procedures which suffer from low recovery and specificity. Detection limit was as low as 2 ppb for furfural by HPLC, while the lowest detection limit of furfural was 75 ppb by colorimetric procedure (*20*). Isocratic HPLC system seems to work for separating HMF and furfural which are structurally very alike (Figure 4). The use of Capillary Electrophoresis (CE), which combines automated chromatography with electrophoresis, to separate Maillard reaction products has received scant attention (*46, 47*). Since these furan compounds (HMF and furfural) are non-ionic in nature, addition of anionic surfactant such as sodium dodecyl sulphate (SDS) was required to separate HMF and furfural. In our lab., we find no advantage of CE over HPLC especially for sensitivity and resolution for HMF and furfural in aged orange juice samples. Since the chemical matrix of citrus juices is complicated, sample clean-up procedures including dilution (*44, 45*), distillation (*20, 41*), and liquid extraction (*43*) prior to HPLC analysis were proposed to remove potential interferences from other constituents, and to maintain the integrity of the analytical column for repetitive use. Although liquid-liquid extraction practices are chosen frequently, solid-phase extraction method are preferred today for convenience, reproducibility and adaptability to automation. In our laboratory we have developed a 2 step procedure including Carrez clarification and C-18 solid-phase extraction for both HMF and furfural in citrus juices (*48*).

The capabilities of modern detectors to monitor at several wavelengths simultaneously and to record spectra of separated peaks adds considerably to analytical capabilities for identification of these in a complex mixture of extracts. The employment of a photodiode array (PDA) detector was of value in demonstrating the spectral capabilities of PDA in separating marker compounds and for tentative identification of HMF (45, 49). Structural elucidation of HMF isolated from orange powder was done previously by Berry and Tatum (28) by IR and UV spectrum and by mass spectrometry (MS). Molecular weight for HMF was 126 and absorption maxima were at 227 nm and 281 nm in a 95% ethanol solution (28). Kim and Richardson (45) collected HMF peak from HPLC analysis of fruit samples and analyzed by GC-MS. The parent molecular weight of 126 and the typical electron impact fragment with m/z of 109, 97 and 69 for HMF were found. Characteristic double maxima (230 and 285 nm) were also found by PDA.

## Furanones

Furanones have a planar enol-oxo-configuration and a caramel-like odor. Among the many products from carbohydrate degradation, 3(2H)-furanones belong to the most striking aroma compounds. The conversion of sugars into furanone seems to involve a series of rearrangement and tautomerization, but whether the furanones detected in citrus products are the favored reaction products at low pH of citrus and are obtained exclusively by a nonenzymic reaction is not yet clear.

**2,5-Dimethyl-4-hydroxy-3(2H)-furanone.** Furaneol (tradename of a 2,5-dimethyl-4-hydroxy-3(2H)-furanone) is an important flavor compound and can be found in various fruits such as pineapple, strawberries, grapes, raspberries, tomato and arctic brambles. Many factors are known to influence natural variations of furaneol contents. Since Furaneol itself has a "burnt pineapple" odor, and is used to enhance the flavor of some fruits products, it may not be appropriate to be considered as a chemical marker compound in such fruit products for quality deterioration. However, to our knowledge, Furaneol was not reported as a natural constituent in citrus, but as one of the degradation products of sugars (50) and is known to be responsible for characteristic pineapple-like note of aged canned orange juice (51). It is likely that Furaneol may be formed through a series of enolization and dehydration reactions of Amadori compounds (52, 53).

From a previous storage test with canned grapefruit juices at temperature between 10 to 50°C, we found that Furaneol content gradually increased as a function of storage time and temperature (42). It was suggested that Furaneol may serve as a quality deterioration index, as well as furfural and HMF in temperature-abused grapefruit juices. Figure 5 shows the result from a storage test with bottled grapefruit juice from 10 to 50°C. Furaneol content gradually increased as a function of storage time and temperature. There was no significant accumulation of Furaneol in juice samples stored at 10 to 20°C (Figure 5). The bottled grapefruit juice stored at 50°C for 15 weeks contained about 25 times more Furaneol than the juice stored at the same period at 20°C. The increase in Furaneol level is highly important to juice quality because it can mask or depress the full orange-like aroma at 0.05 ppm level (*51*) and is reported as a powerful flavoring agent with odor threshold value of 0.1-0.2 ppm level in water (*54*).

Furaneol is a thermally unstable compound, both easily formed and readily degraded (*55*). This furanone may interact further with amino acids and/or other degradative products to form various secondary flavor compounds (*56*). Shaw and coworkers (*55*) presented effects of pH and temperature on thermal generation of Furaneol in model systems using the reaction of rhamnose and proline. Rhamnose and proline are known to be precursors of many pleasant aromas which are important to the food industry. The content of thermally generated Furaneol roughly follows the concentration of rhamnose. Rhamnose is a minor constituent of citrus juices, being derived by enzymatic degradation of pectin during citrus juice processing (*57*).

Figure 6 shows the detailed information about stability of Furaneol in aqueous buffer solutions (pH 2-8) and at temperatures between 5 and 100°C (*58*). The decomposition of Furaneol was known to follow first-order kinetics and was dependent on the pH of the solution. The stability of furaneol had an optimum at pH 4.0 (Figure 6a) and the half-life of Furaneol at ambient temperature and pH 4.0 was 100 days (Figure 6b). The decomposition of Furaneol increased very rapidly with rising temperature. At 100°C, the decomposition of Furaneol proceeded about 200 times faster than at 20°C (Figure 6c and 6d). Furaneol can be a good indication 2,3-enolization pathway of sugar degradation in citrus. The main pathway of sugar degradation at the acidic pH of citrus juice is considered to be 1,2-enolization, but it is also natural that 2,3-enolization occurs to some extent (*53*). Under the same storage conditions, significantly small amount of Furaneol compared to HMF (ca. 200 times) or furfural (ca. 14 times) was observed in aged bottled grapefruit juices as shown in Figure 7.

Figure 8 shows structures of some potential quality indicators for citrus products. Analysis of this oxygen-containing heterocyclic Furaneol (Figure 8) is somewhat difficult, since it is unstable in air and in an aqueous solution and can not be steam distilled due to being highly oxygenated in nature. Thus, a sample clean up procedure using solid-phase extraction with C-18 Sep-Pak cartridge was developed in our laboratory to provide reproducible results (*42*). The mean recovery of C-18 solid-phase extraction with ethyl acetate was 92.6% with 2.5% CV. RP-HPLC on a ODS-A, S-5 column (250 X 4.6 mm, id, 120 Å, YMC Inc., Wilmington, NC) with mobile phase of 0.05M sodium acetate (pH 4.0) : acetonitrile (80:20, v/v), UV detection at 290nm, at 30°C could detect this unstable compound as low as 50 ppb without deterioration (Figure 9). Recently, a C-18 solid phase extraction procedure to isolate both 4-vinyl guaiacol and Furaneol from stored orange juice in a single extraction was proposed (*59*). GC analysis involving methylene chloride extraction was also described for furanones in mango (*60*) and in tomato (*61*). GC-MS, electrospray tandem mass spectrometry (ES-MS/MS) and NMR techniques were also applied for positive identification of furanones (*61*).

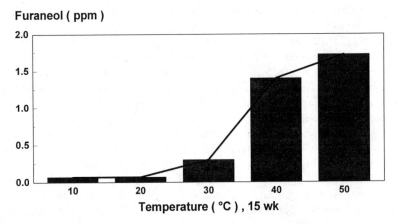

Figure 5. Effects of storage temperature on the formation of furaneol in glass bottled grapefruit juice.

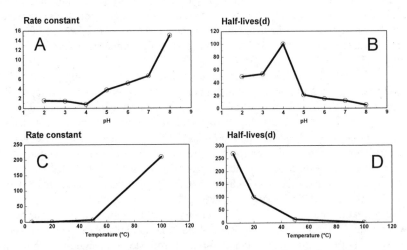

Figure 6. Effects of temperature and pH on Furaneol (0.1mM) stability. Conditions of (A) & ( B ) was 20 °C, and (C) & (D) at pH 4.0. Adapted from Ref. 58.

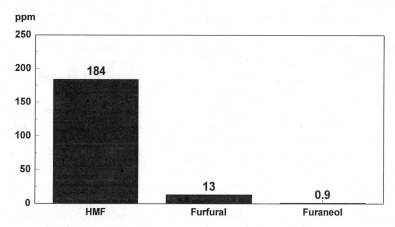

Figure 7. Comparative formation of HMF, furfural and Furaneol in bottled grapefruit juice stored  after 12 wks at 50°C.

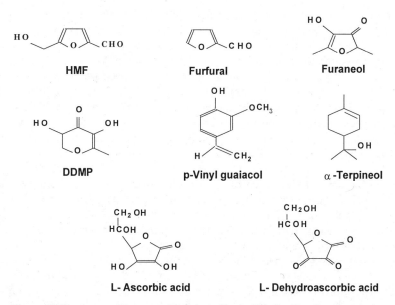

Figure 8. Structures of some potential quality indicators for citrus products.

## Pyranones

Pyranones are oxygen-containing heterocyclic compounds associated with caramel-like flavor notes and are derived from thermal degradation of carbohydrates.

**2,3-Dihydro-3,5-dihydroxy-6-methyl-4H-pyran-4-one.** 2,3-dihydro-3,5-dihydroxy-6-methyl-4H-pyran-4-one (DDMP) was isolated from acid-catalyzed fructose dehydration products (*62, 63*) and from dehydrated orange juice (*64*). Recently, this compound has been identified by GC-MS, LC-MS and UV in citrus product by Kim and Taub (*49*) and suggested as one of the marker compounds in various heated foods including orange juice. The parent MW is 144 by LC-MS (*49*) and UV maximum is 298nm (*62*). This compound can be produced from fructose, by Maillard reaction or by a combination of both and is known to increase to a limiting value in broccoli (*49*). Earlier, Askar (*26*) indicated that this pyranone is a product typical for the Maillard reaction in fruit juices and undergoes acid-catalyzed rearrangement to yield isomaltol, a furan derivative (*63*). This pyranone has a high threshold value, is not available from a commercial source, and no kinetic data and detection method have been established in citrus products as yet.

## Phenols

Hydroxycinnamic acids (HCAs), comprising p-coumaric, ferulic, caffeic and sinapic, and their bound forms, are found in citrus fruit parts (*65*). During processing and storage of citrus juices, vinyl phenols are produced from the free HCA by acid-catalyzed decarboxylation. The decarboxylation of all HCAs would potentially produce p-vinyl phenol (from p-coumaric acid), p-vinyl guaiacol (ferulic acid), p-vinyl catechol (caffeic acid) and 3,5-dimethyl-4-hydroxystyrene (sinapic acid). P-vinyl phenol and p-vinyl guaiacol have been identified in processed citrus juices, but p-vinyl catechol and 3,5-dimethyl-4-hydroxy styrene have yet to be reported. Vinyl phenols are unpleasant smelling compounds with very low perception thresholds; their presence adversely affects acceptability of citrus juice products.

**4-Vinyl Guaiacol.** 4-vinyl guaiacol has been identified as the most detrimental off-flavor compound in aged canned orange juice (*51*). This compound imparts an old fruit or rotten flavor to orange juice at a level of 0.075 ppm. Also, this compound is responsible for an undesirable phenolic flavor in beer (*66*). In spirits, since phenols impart a distinctive aroma to whiskeys, the qualitative determination of simple phenols in whisky has been used to give a phenol fingerprint.

    4-vinyl guaiacol was the primary product from thermal decarboxylation of ferulic acid (*67*). In orange juice, formation of 4-vinyl guaiacol has been confirmed using the model systems of orange juice with addition of ferulic acid (*68, 69*). In a comparative study of decarboxylation of cinnamic acids in the pH range 1-6 at 100°C, the substituent (hydroxy or methoxy) and position of the substituent was found to affect the rate of decarboxylation (*70*). The maximum in the rates of decarboxylation of ferulic acid was observed at about pH 5 . The kinetics of thermal decomposition of ferulic acid was evaluated by using the nonisothermal method to determine the order of reaction and the activation energy (*67*). Since the rate is accelerated by oxygen, decarboxylation is probably a radical reaction.

Detection of 4-vinyl guaiacol can be a good indication of decarboxylation of phenolic compounds in citrus juice during thermal processing and subsequent storage, and can provide further understanding of its influence on off-flavor in juice. However, the effectiveness of 4-vinyl guaiacol as a chemical marker to assess quality deterioration is not encouraging. From a previous study with a model system, Naim et al. (*68*) indicated that only trace amounts of 4-vinyl guaiacol could be found in model orange juice samples even though the degradation of ferulic acid with time and accelerated nonenzymic browning were evident. Also, 4-vinyl guaiacol seems to be formed and then degraded significantly during prolonged storage at high temperatures. From storage test with single-strength citrus juices (*71*), we observed that the concentration of 4-vinyl guaiacol increased with storage time at all temperatures; however, a noticeable decrease occurred after extended storage at 50 °C as shown in Figure 10. It appears that ferulic acid can be decomposed to 4-vinyl guaiacol, which in turn appears to be the source of other phenolic products (*67*) or is transformed into its dimer and trimer (*72*). As a result, it will be difficult to characterize the kinetics of quality deterioration based on 4- vinyl guaiacol contents.

Previous work with storage-aged orange juice with a conventional UV detector did not detect any formation of 4-vinyl guaiacol at low temperature below 30°C. Since 4-vinyl guaiacol has native fluorescence (maximum excitation at 320nm and emission at 353nm), an HPLC method using fluorescence detection was proposed (*71*). Fluorescence detection possesses sufficient sensitivity to detect as low as 10 ppb for 4-vinyl guaiacol and can detect as low as one fifth the flavor threshold value reported in literature in orange juice (*51*). Juice sample was prepared using C-18 solid-phase extraction with recovery rate of 91.2%. Binary gradient elution with acidified aqueous acetonitrile was used for elution of 4-vinyl guaiacol on C-18 column. This method could overcome inconvenience of TLC method for 4-vinyl guaiacol in orange juice (*68*).

## Oil Components

The presence of essential oils in citrus juices is responsible for imparting pleasant aromas and enhancing the taste of these products. However, under unfavorable processing, storage and packaging conditions, some essential oil components degrade to yield objectionable off flavors. These series of detrimental reactions proceed by way of acid catalysis in a dilute aqueous medium and involve hydration, dehydration, rearrangements, cyclizations and hydrolyses (*73, 74*). The constituent oxygen in the newly formed oxygen-containing compounds is derived from water by chemical addition, and not from dissolved oxygen by oxidation. Table I summarizes thermally degraded compounds derived from constituents.

**α-Terpineol.** Off-flavor notes derived from some essential oil compounds have been described as terpeney (terpentine-like), camphoraceous, stale, musty and pungent. As noted by several storage studies, some thermally degraded oil constituents are amenable in monitoring quality changes. During storage of aseptically packaged

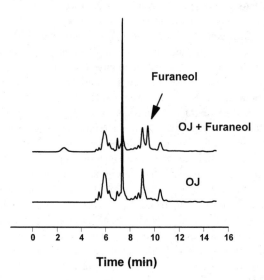

Figure 9. HPLC chromatograms of Furaneol. Conditions are in the text.

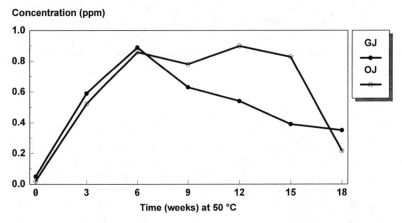

Figure 10. Changes of 4-vinyl guaiacol contents in grapefruit (●) and orange (O) juice at 50°C.

Table I. Thermally degraded compounds derived from essential oil constituents

| Precursor | Derived Compounds | Flavor Response |
|---|---|---|
| *d*-limonene | α-terpineol | stale, musty, piney |
| linalool | α-terpineol<br>nerol<br>geraniol | stale, musty, piney<br>sweet, rose fruity<br>sweet, floral rose |
| α-terpineol | *cis*-1,8-menthanediol | sweet, camphoraceous |
| *cis*-1,8-*p*-menthanediol[a] | 1,8-cineol<br>1,4-cineole | pungent, camphoraceous |
| citral[b] | *p*-mentha-1,5-dien-8-ol<br>*p*-mentha-1(7), 2-dien-8-ol<br>*cis*-*p*-mentha-2,8-dien-1-ol<br>*trans*-*p*-mentha-2,8-dien-1-ol | not characterized<br>not characterized<br>not characterized<br>not characterized |
| *p*-mentha-1,5-dien-8-ol | *p*-cymen-8-ol | nonspecific off-flavor |
| *p*-mentha-1(7),2-dien-8-ol | *p*-cymene<br>α, *p*-dimethylstyrene | terpeney off-flavor<br>terpeney off-flavor |
| ν- terpinene | *p*-cymene | terpeney off-flavor |

[a]Also known as *cis*-terpinol
[b]Isomeric mixture of neral and geranial
Adapted from Ref. 81.

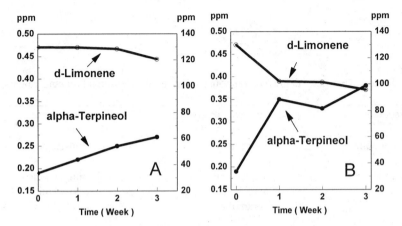

Figure 11. Changes of d-limonene (O) and α-terpineol (●) contents in glass (A) and EVOH-barrier packed (B) orange juice at 1.1°C. Adapted from Ref. 77.

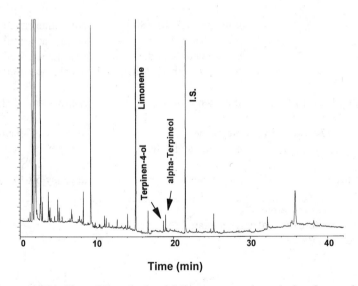

Figure 12. Capillary GC analysis of d-limonene, terpinen-4-ol and α-terpineol in orange juice. Internal standard is methyl decanoate.

orange juice, Durr et al. (*75*) noted significant decreases in limonene, linalool, neral, geranial, octanal and decanal, and an increase in one undesirable component, α-terpineol. Tatum and coworkers (*51*) showed that as little as 2.5 ppm α-terpineol added to freshly expressed juice caused a stale, musty or piney note. Moshonas and Shaw (*76*) monitored 29 volatile components of aseptically packaged orange juice during 8 months of storage. A gradual decrease occurred in several flavor components, namely, 1-penten-3-one, hexanal, ethyl butyrate, octanal, neral and geranial, whereas α-terpineol and furfural increased. Moshonas and Shaw (*76*) stated that their flavor panel considered the stored juice to have an aged or stale flavor rather than a heated or processed flavor. Development of a stale flavor in orange juice has been attributed to the loss of volatile esters (*73*) and to the buildup of α-terpineol during storage (*51*). Data in Figure 11a and 11b shows the more rapid increase of α-terpineol concentration in the orange juice packed with EVOH-barrier compared to the same juice stored in glass over a 3 week period at 1.1°C (*77*). The increase of α-terpineol was also known to be much more dependent on the storage temperature rather than on the initial limonene content of the juice (*75*).

Numerous GC studies have been conducted for analysis of citrus flavor and aroma components in order to obtain accurate quantitation of losses of desirable flavor components and the appearance of undesirable limonene degradation products during processing and storage. Most GC procedures involved simultaneous distillation-solvent extraction (*78*), vacuum distillation (*79*), or simple pentane-extraction technique (*77*). Figure 12 shows the detection of α-terpineol from many volatile components in orange juice by capillary GC analysis in our laboratory. A 10mL of sample of canned orange juice was centrifuged and 2 mL of the clarified serum was passed through a C-8 Sep-Pak cartridge. The components were eluted from the cartridge with ethyl ether-pentane, concentrated to a small volume, and resolved by capillary gas chromatography (*80*).

**Literature Cited**

1.   Kefford, J. F.; McKenzie, H. A.; Thompson, P. C. O. *J. Sci. Food Agric.* **1958**, 10, 51-63.
2.   Huelin, F. E. *Food Res.* **1953**, 18, 633-639.
3.   Smoot, J. M.; Nagy, S. *J. Agric. Food Chem.* **1980**, 28, 417-421.
4.   Nagy, S.; Smoot, J. M. *J. Agric. Food Chem.* **1977**, 25, 135-138.
5.   Johnson, J .R.; Braddock, R. J.; Chen, C. S. *J. Food Sci.* **1995**, 60, 502-505.
6.   Lee, H. S.; Johnson, J. R.; Barros, S. M.; Chen, C. S. In *Book of abstracts-1994 IFT Annual Meeting Technical Program*, #22-6, IFT,: Atlanta, GA. **1994**, pp. 54.
7.   Kurata, T.; Fujimaki, M.; Sakurai, Y. *Agric. Biol. Chem.* **1973**, 37, 1471-1477.
8.   Sawamura, M.; Takemoto, K.; Li, Z.F. *J. Agric. Food Chem.* **1991**, 39, 1735-1737.
9.   Sawamura, M.; Takemoto, K.; Matsuzaki, Y.; Ukeda, H.; Kusunose, H. *J. Agric. Food Chem.* **1994**, 42, 1200-1203.

10.    Li, Z. F.; Sawamura, M.; Yano, H. *Agric. Biol. Chem.* **1989**, 53, 1979-1981.
11.    Hoare, M.; Lindsay, J. *J. Food Australia* **1993**, 45, 341-345.
12.    Dennison, D. B. ; Kirk, J. R. *J. Food Sci.* **1978**, 43, 609-612 & 618.
13.    Bui-Nguyen, M. A. In *Modern chromatographic analysis of the vitamins,* DeLeenheer, A. P.; Lambert, W. E.; DeRuyter, M. G. M., Eds.; Marcel Dekker, Inc.: New York, NY, **1985**. pp. 267-301.
14.    Goldenbergm, H.; Jirivetz, L.; Krajnik, P.; Mosgoller, W.; Moslinger, T.; Schweinzer, E. *Anal. Chem.* **1994**, 66, 1086-1089.
15.    Kim, H.J. *J. Assoc. Off. Anal. Chem.* **1989**, 72, 681-686.
16.    Lee, H. S.; Nagy, S. *J. Food Sci.* **1988**, 53, 168-172 & 180.
17.    Lee, H. S.; Nagy, S. *Food Technol*, **1988**, 42, 91-97.
18.    Nagy, S. ; Randall, V. *J. Agric. Food Chem.* **1973**, 21, 272-275.
19.    Maraulja, M. D.; Blair, J. S.; Olsen, R. W. ; Wenzel, F.W. *Proc. Fla. State Hort. Soc.* **1973**, 86, 270-275.
20.    Marcy, J.; Rouseff, R .L. *J. Agric. Food Chem.* **1984**, 32, 979-981.
21.    Fors, S. In *The Maillard reaction in foods and nutrition,* Waller, G.R.; Feather, M.S. Ed.; Am. Chem. Soc.: Washington, DC. **1983**, pp.185-286.
22.    Kanner, J.; Harel, S.; Fishbein, J.; Shalom, P. *J. Agric. Food Chem.* **1981**, 29, 948-949.
23.    Kanner, J.; Fishbein, J.; Shalom, P.; Harel, S.; Ben-Gera, I. *J. Food Sci.* **1982**, 47, 429-431.
24.    Herrmann, J. ; Partassidou, V. *Die Nahrung*, **1979**, 23, 143-150.
25.    Singh, B.; Dean, G. R.; Cantor, S. M. *J. Am. Chem. Soc.* **1948**, 70, 517-522.
26.    Askar, A. *Fluss. Obst.* **1984**, 51, 564-569.
27.    Koch, V. J. *Deutsche Lebensm. Rund.* **1966**, 62, 105-108.
28.    Berry, R. E. ; Tatum, J. H. *J. Agric. Food Chem.* **1965**, 13, 588-590.
29.    Wucherpfennig, K.; Burkardt, D. *Fluss. Obst.* **1983**, 50, 416- 422.
30.    Lee, H. S.; Nagy, S. *J. Food Process. Preserv.* **1990**, 14, 171-178.
31.    Toribio, J. L.; Lozano, J. E. *Lebensm. Wiss. Technol.* **1987**, 20, 59-63.
32.    Shallenberger, R. S.; Mattick, L. R. *Food Chem.* **1983**, 12, 159-165.
33.    Resnik, S. ; Chirife, J. *J. Food Sci.* **1979**, 44, 601-605.
34.    Lozano, J. E. *Lebensm. Wiss. Technol.* **1991**, 24, 355-360.
35.    Yaylayan, V. ; Forage, N. G. *J. Agric. Food Chem.* **1991**, 39, 364-369.
36.    *RSK-Values, The complete manual.* Bielig, H.J.; Faethe, W.; Fuchs, G.; Koch, J.; Wallrauch,S.; Wucherpfennig, K., Eds.; Verlag Flussiges Obst. GmbH,: Schonborn, Germany, **1987**, pp.15-32.
37.    Meydav, S. ; Berk, Z. *J. Agric. Food Chem.* **1978**, 26, 282-285.
38.    Berk, Z.; Mannheim, C. H. *J. Food Process. Preserv.* **1986**, 10, 281-293.
39.    Winkler, O. *Z. Lebensm-Untersuch-Forsch.* **1955**, 102, 161-172.
40.    Cilliers, J. L. ; Van Niekerk, P.J. *J. Assoc. Off. Anal. Chem.* **1984**, 67, 1037-1039.
41.    Dinsmore, H. L. ; Nagy, S. *J. Assoc. Off. Anal. Chem.* **1974**, 57, 332-335.

42.  Lee, H. S.; Nagy, S. *J. Food Sci.* **1987**, 52, 163-165.
43.  Mijares, R.; Park, G. L.; Nelson, D. B.; McIver, R. C. *J. Food Sci.* **1986**, 51, 843-844.
44.  Li, Z. F.; Sawamura, M.; Kusunose, H. *Agric. Biol. Chem.* **1988**, 52, 2231-2234.
45.  Kim, H. J. ; Richardson, M. *J.Chromatogr.* **1992**, 593, 153-156.
46.  Corradini, D.; Corradini, C. *J. Chromatogr.* **1992**, 624, 503-509.
47.  Tomlinson, A. J.; Landers, J. P.; Lewis, I. A. S.; Naylor, S. *J. Chromatogr.* **1993**, 652, 171-177.
48.  Lee, H. S.; Rouseff, R. L.; Nagy, S. *J. Food Sci.* **1986**, 51, 1075-1076.
49.  Kim, H. J.; Taub, I. A. *Food Technol.* **1993**, 47, 91-99.
50.  Shaw, P. E.; Tatum, J. H.; Berry, R. E. *J. Agric. Food Chem.* **1968**, 16, 979-982.
51.  Tatum, J. H.; Nagy, S.; Berry, R. E. *J. Food Sci.* **1975**, 40, 707-709.
52.  Baltes, W. *Food Chem.* **1982**, 9, 59-73.
53.  Hodge. J. E.; Fisher, B. E.; Nelson, E. C. *Am. Soc. Brewing Chem.* **1963**, 1963, 84-92.
54.  Pittet, A. O.; Rittersbacher, P.; Muralidhara, R. *J. Agric. Food Chem.***1970**, 18. 929-933.
55.  Shaw, J. J.; Burris, D.; Ho, C.T. *Perfumer & Flavorist*, **1990**, 15, 60-66.
56.  Shu, C. K.; Mookherjee, B. D.; Ho, C. T. *J. Agric. Food Chem.* **1985**. 33. 446-448.
57.  Handwerk, R. L.; Coleman, R. L. *J. Agric. Food Chem.* **1988**, 36, 231-236.
58.  Hirvi, T.; Honkanen, E.; Pyysalo, T. *Lebensm. Wiss. Techol.* **1980**, 13, 324-325.
59.  Walsh, M.; Rouseff, R. L.; Naim, M.; Zehavi, U. In *Book of abstracts - 1995 IFT Ann. Meeting Technical Program,* #26D-19, IFT, :Anaheim, CA. **1995**, pp. 74.
60.  Wilson C. W.; Shaw, P.E.; Knight, R.J. *J. Agric. Food Chem.* **1990**, 38, 1556-1559.
61.  Krammer, G. E.; Takeoka, G. R.; Buttery, R. G. *J. Agric. Food  Chem.* **1994**, 42, 1595-1597.
62.  Shaw, P. E.; Tatum, J. H.; Berry, R. E. *Carbohydr. Res.* **1967**, 5, 266-273.
63.  Shaw, P. E.; Tatum, J. H.; Berry, R. E. *Carbohydr. Res.* **1971**, 16, 207-211.
64.  Shaw, P. E.; Tatum, J. H.; Berry, R. E. *Dev. Food Carbohydr.* **1977**, 1, 91-111.
65.  Naim, M.; Zehavi, U.; Nagy, S.; Rouseff, R. L. In *Phenolic compounds in food and their effects on health,* vol. 1. Ho, C.T.; Lee , C.Y.; Huang, M.T. Eds. ; Am. Chem. Soc.: **1992**, pp. 180-191.
66.  Maga, J. A. *CRC Critic. Rev. Food Sci. Nutr.* **1978**, 10, 323-372.
67.  Fiddler, W.; Parker, W. E.; Wasserman, A. E.; Doerr, R. C. *J. Agric. Food Chem.* **1967**, 15, 757-761.
68.  Naim, M.; Striem, B.J.; Kanner, J.; Peleg, H. *J. Food Sci.* **1988**, 53, 500-503 & 512.

69.    Peleg, H.; Striem, B. J.;  Naim, M.; Zehavi, U. *Proc. Int. Soc. Citricult.* **1988**, 4, 1743-1748.
70.    Pyysalo, T.; Torkkeli, H.; Honkanen, E. *Lebensm. Wiss. Technol.* **1977**, 10, 145-147.
71.    Lee, H. S. ; Nagy, S. *J. Food Sci.* **1990**, 55, 162-163 & 166.
72.    Klaren-De Wit, M.; Frost, D. J.; Ward, J. P. *Recl. Trav. Chim. Pays-Bas*, **1971**, 90, 906-911.
73.    Blair, J. S.; Dodar, E. M.; Masters, J. E.; Riester, D. W. *Food Res.* **1952**, 17, 235-260.
74.    Clark, B. C. ;  Chamblee, T. S. In *Off-flavors in foods and beverages*, Charalambous, G. Ed.; Elsevier Science Pub.: New York, NY, **1992**, pp. 229-285.
75.    Durr, P.; Schobinger, U.;  Maldvogel, R. *Alimenta*. **1981**, 20, 91-93.
76.    Moshonas, M. G.; Shaw, P. E. *J. Agric. Food Chem.* **1989**, 37, 157-161.
77.    Marsili, R.;  Kilmer, G.; Miller, N. *LC-GC*. **1989**, 7, 779-783.
78.    Nunez, A. *J. Chromatographia*. **1984**, 18, 153-158.
79.    Moshonas, M. G.; Shaw, P. E. *J. Agric. Food Chem.* **1984**, 35, 161-165.
80.    Nagy, S.;  Chen, C. S.; Barros, S.; Carter, R. *Proc. Fla. State Hort.Soc.* **1990**, 103, 272-275.
81.    Nagy, S.; Rouseff, R. L.; Lee, H. S. In *Thermal generation of aromas*, Parliment, T. H.; McGorrin, R. J.; Ho, C. T., Eds.; Am. Chem. Soc.: Washington, D.C. **1989**, pp.331-345.

Chapter 10

# Shelf-Life Prediction of Aseptically Packaged Orange Juice

L. M. Ahrné[1], M. C. Manso[2], E. Shah[1], F. A. R. Oliveira[2],
and R. E. Öste[3]

[1]Tetra Pak Processing Systems Division, Rubens Rausing gata,
S–221 86 Lund, Sweden
[2]Escola Superior de Biotechnologia, Universidade Católica Portuguesa,
R. Dr. António, Bernardino de Almeida, 4200 Porto, Portugal
[3]Department of Applied Nutrition and Food Chemistry, Chemical Centre,
University of Lund, P.O. Box 124, S–221 00 Lund, Sweden

The prediction of shelf-life is a useful tool for the development of
new products and packaging materials. The prediction of product
shelf-life can be obtained by mathematical modelling of the
changes in selected quality characteristics of packaged food during
storage. In order to investigate the quality changes during storage,
orange juice with five different initial oxygen concentrations (1 to
10 ppm) was filled in 1 L Tetra Brik Aseptic cartons, and stored at
seven different temperatures (50, 40, 30, 20, 8, 4 and -18°C).
Samples of orange juice were collected during time course and
were analysed for ascorbic acid content, browning index, off-
flavour production (p-vinyl-guaiacol). Mathematical models
including the effect of the storage temperature and the initial
concentration of dissolved oxygen in the juice were developed for
the quality characteristics analysed. The models showed to be
adequate to describe the changes in orange juice quality during
storage.

The world trade in fruit juices has trebled since 1980, and there is good reason to
believe that it will remain a growth industry for a long time to come. The average
per capita consumption of fruit juices and fruit drinks increased from 10 L/year to
45 L/year in Western Europe, partly due to greater consumer awareness which is a
good selling point for juices. Aseptically packaged fruit juice that does not need
refrigeration is well established in Europe, and is becoming more popular in the
USA.
    The shelf-life of a food is the time period for the product to become
unacceptable from sensorial, nutritional or safety perspectives (1). In this way,
shelf-life is a measure of quality changes during storage. The main changes in
orange juice quality during storage are caused by microbial growth and chemical
reactions, which are accelerated by the environment factors. The environmental
factors, such as temperature, relative humidity, light and oxygen will define shelf-

0097–6156/96/0631–0107$15.00/0

life of orange juice. The initial quality of the juice and the processing conditions will also influence the shelf-life of the packaged juice. The product shelf-life can be predicted by mathematically modelling the quality changes. A more general model can be obtained if the influence of external factors is included in the model. Singh and Heldman (2) developed a computer programme to predict the ascorbic acid degradation of an infant formula. The influence of factors such as the initial dissolved oxygen, light intensity and oxygen permeability of container wall were also considered in the model. No references were found in literature, where the quality losses of aseptically packaged juices during storage and the influence of external factors were mathematically modelled.

To establish mathematical models for the quality change during storage and to predict the shelf-life of packaged product, it is necessary:

- to identify and quantify the orange juice quality characteristics that significantly change during storage;
- to identify the factors that influence the quality change and
- to develop mathematical models to describe the deterioration of quality and influence of the external factors.

In aseptically packed orange juice, microbial growth and enzymatic reactions are inhibited. Thus, the main change in juice quality is caused by nonenzymatic chemical reactions. These reactions adversely affect the nutritional value, flavour and colour of orange juice. Orange juice quality in terms of nutritional value is best quantified by the amount of ascorbic acid. However, it is well known that ascorbic acid degradation promotes browning and flavour changes in the orange juice.

The storage temperature, the oxygen initially dissolved in the juice and the oxygen that permeates through the package are considered to be the major factors that reduce shelf-life of aseptically packaged juices.

The objective of this work is to develop general models that will predict the ascorbic acid loss, browning development and formation of p-vinyl-guaiacol (PVG) in packaged orange juice, with different oxygen levels, during storage from 4 to 50°C.

**Materials and Methods**

**Processing and Packaging.** Single Strength Valencia Orange Juice (12° Brix) was obtained from Cargill Citro-America, Inc. In the Tetra Pak pilot plant (Lund, Sweden) the juice was aerated or deaerated in order to obtained different dissolved oxygen concentrations and then pasteurised at 95°C for 15 sec. Orange juice with initial oxygen concentrations of 10, 6, 2, 1.4 or 1.0 ppm was aseptically packaged in one litre Tetra Brik Asept (TBA) cartons. For each oxygen concentration, a total of 140 TBA were filled. The cartons were equally distributed into seven storage rooms held at -18, 4, 8, 20, 30, 40, 50°C. Ten TBA cartons (two of each dissolved oxygen concentration) were collected from each storage room at different time intervals depending on temperature.

**Ascorbic Acid Concentration.** An enzymatic kit from Boeringer Mannheim (Sweden) was used for determination of ascorbic acid concentration in the juice. Before measurement, the packages were shaken and 1 mL orange juice was collected in an eppendorf tube. The juice was then centrifuged at 4000 rpm for 20 minutes and the supernatant was used for analysis as described by the manufacturer (Boeringer cat. No 409677).

**Browning Index.** Browning index was measured as the absorbance at 420 nm (*3*) using a DMS 100 UV-Visible spectrophotometer (Varian, Sweden). The brown pigments were extracted by dilution of the juice 1:1 with 95% ethyl alcohol. Before measuring the juice was filtered through a Munktell 00H analytical filter paper.

**Off-flavour Determination.** The p-vinyl-guaiacol was determined by HPLC using the method described by Lee and Nagy (*4*). Before HPLC analysis the juice samples were centrifuged for 10 minutes at 12,000 x g. One millilitre of supernatant was injected into a C-18 Sep-Pak cartridge (Varian, Sweden), which had been preconditioned with 2 mL of methanol followed by 3 mL of water. The 4-vinyl-guaiacol was eluted with 0.3 mL of acetronitrile followed by 0.7 mL of water. The eluted sample was filtered through a 0.45 μm Millipore filter (Millipore Corporation, Milford, MA, USA) before injection. The vinyl guaiacol used to prepare the standard solutions was purchased from Lancaster Synthesis (UK).

The samples were analysed by an HPLC equipped with a LiChrospher 100 RP-18 column (5 mL, 250 mm x 4 mm, Merck, Sweden) with a RP-18 precolumn (25 mm x 4 mm, Merck, Sweden). A linear gradient and isocratic elution with 1% acetic acid in water (solvent A) and 1% acetic acid in acetonitrile (solvent B) was used with a flow rate of 1 mL/min: 0-20 min, linear gradient, 90 to 68% of A, 10 to 32% of B; 20-25 minutes, isocratic, 68 % of A; 25-35 minutes, linear gradient, 68 to 58 % of A, 32 to 42% of B. The column was washed with 100% B for 10 minutes, and equilibrated with 90% A- 10% B for 10 minutes. Analysis was carried out at room temperature by injecting 80 mL of the sample or standard into the column. Each sample was injected twice. The separated chromatographic peaks were identified and integrated with a Photo Diode Array detector (Varian, Sweden).

## Mathematical Modelling

The effect of temperature and initial concentration of dissolved oxygen was included in the kinetic models using the systematic procedure described in Figure 1. The data of the quality indicators, ascorbic acid concentration, browning development and p-vinyl-guaiacol formation were obtained for the different temperatures during the storage time. These data were then fit to adequate kinetic equation. The parameters of the kinetic equations were obtained by the best fit for the experimental data. These parameters were obtained at all temperatures and for all the batches with different oxygen concentrations.

The next step was to assess the dependence of these parameters on temperature. In kinetic models, it is usually assumed that the rate constant varies with temperature according to the Arrhenius equation. Deviations to this equation have been reported by several authors (*5, 6*). When a wide temperature range is studied, as in this work, changes in reaction mechanisms may be observed. In this case the dependence of the rate constant with temperature was described by two Arrhenius equations. If one or two straight lines were identified in the plot of ln k versus 1/T, the dependence of the rate constant on temperature was described by one or two Arrhenius equations. The parameters of the Arrhenius equation, i.e. the activation energy and the pre-exponential factor, were obtained for all the batches with different initial concentration of dissolved oxygen. Whenever a variation of parameters of the Arrhenius equation with oxygen was observed, its effect was included in the model by expressing the variation of the activation energy and the logarithm of the pre-exponential factor with the amount of dissolved oxygen by a linear equation.

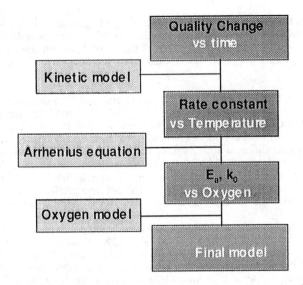

Figure 1. Schematic representation of the procedure used to develop the mathematical models.

The final step of this procedure was to check the accuracy of the model, which was done by comparing predicted and experimental values. The error in the prediction was calculated for each storage time, as % error = (experimental value - predicted value) x 100 /experimental value.

## Results and Discussion

**Ascorbic acid.** The ascorbic acid loss during storage showed an initial period of rapid loss followed by a period of slow loss. This behaviour can be described by a two simultaneous first order kinetic model. This means that two simultaneous reactions with different rates are responsible for the ascorbic acid degradation. It has been referred in literature that ascorbic acid is degraded by two simultaneous reactions: one aerobic and other one anaerobic. The aerobic reaction dominates first and it is fairly rapid, while the anaerobic reaction dominates later and it is quite slow (6,7). Equation (1) was used to fit the experimental data.

$$C = C_0 \exp(-k_0 t) + C_1 \exp(-k_1 t) \tag{1}$$

In this equation:
    $C$ (ppm) is the amount of ascorbic acid in the juice at time $t$ (days)

    $C_0$ (ppm) and $k_0$ (days$^{-1}$) represent, respectively, the hypothetical initial amount of ascorbic acid and the rate constant associated with the fast reaction, and

    $C_1$ (ppm) and $k_1$ (days$^{-1}$) represent, respectively, the hypothetical initial amount of ascorbic acid and the rate constant associated with the slow reaction - probably the anaerobic degradation.

The rate constants of the two reactions varied with temperature by an Arrhenius equation. In Figure 2 it can be observed that the fast reaction showed only one Arrhenius profile, while the slow reaction, showed two distinct Arrhenius profiles, one between 4 and 20°C and other between 30 and 50°C. Similar distinct profiles were reported in literature (6). Activation energy values between 56 and 46 kJ/mole were obtained for the fast reaction, respectively for initial concentration of dissolved oxygen of 1 and 10 ppm. A linear dependence of the parameters of the Arrhenius equation on oxygen was observed. For the slow reaction, the activation energy at low temperatures (15 kJ/mole) was 6 times lower than at high temperatures (91 kJ/mole), which results in a much faster degradation at high temperatures than at lower temperatures. The rate constant for this reaction did not change significantly with the initial concentration of dissolved oxygen. This fact confirms the idea that this reaction is associated with the anaerobic reaction.

Figure 3 shows, for an orange juice with an initial concentration of dissolved oxygen of 10 ppm, a good agreement between experimental and predicted values of ascorbic acid during storage at temperatures from 4 to 50°C. Similar results were obtained for the batches with lower amount of dissolved oxygen. As an example of how the adequacy of the models were checked, Figure 4 represents the % error at each temperature and for each storage time.

**Browning Index.** The non-enzymatic browning was modelled by a pseudo zero-order kinetic reaction. This linear model may not be the most adequate to describe the complex group of reactions involved in the orange juice browning at high temperatures. Nevertheless, it is a simple model and proved to be effective to predict the browning of orange juice at all temperatures. In the equation (2),

$$B = B_0 + k_b t \qquad (2)$$

B (abs 420 nm) represents the browning index at time t (days),
$B_0$ represents the initial browning index and
$k_b$ the rate constant for browning reactions (days$^{-1}$)

The value of $B_0$ was considered constant and its value was determined for each batch using 20 experimental values obtained with the control samples.
The value of the rate constant of browning reactions varied with temperature by an Arrhenius type equation in the range of temperature from 20°C to 50°C (Figure 5). Activation energies between 122 and 100 kJ/mole were obtained, respectively, for juice with 1 and 10 ppm. The parameters of the Arrhenius equation vary with the amount of oxygen dissolved by a linear equation. In Figure 5, it can be observed that the effect of temperature is more important than the oxygen, and the effect of oxygen is mainly noticeable at low storage temperatures. Figure 6 shows an example of fit between the predicted and experimental values for the browning, in this case for a juice with an initial concentration of dissolved oxygen of 6 ppm. Similar plots were obtained for juice with other amounts of dissolved oxygen. In this figure it can be also noticed that at temperatures lower than 30°C, no significant changes in the colour of orange juice occurred, while at high temperature in a short storage time the juice becomes quite brown, as expected because of the magnitude of the activation energy values.

**p-Vinyl-Guaiacol Formation.** The last model developed was for the formation of PVG, the indicator of off-flavour production. The PVG was identified by Tantum and coworkers (8) as one of the most malodorous off-flavour compounds in aged canned orange juice. PVG increases significantly during storage of orange juice

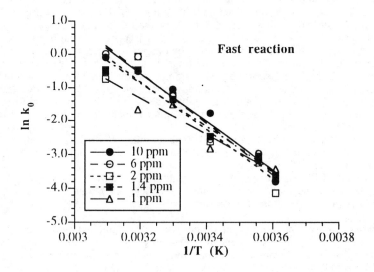

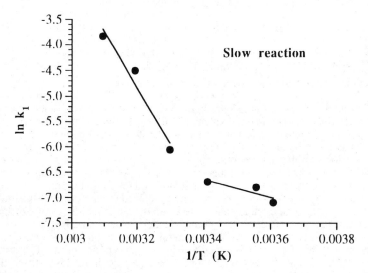

Figure 2. Arrhenius plots for the rate constants ($k_0$ and $k_1$) of ascorbic acid degradation

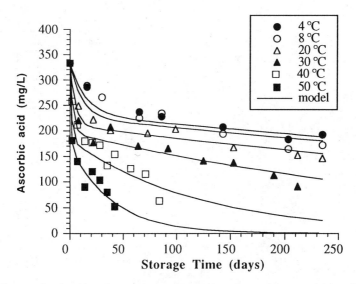

Figure 3. Predicted and experimental values of ascorbic acid concentration during storage at 4, 8, 20, 30, 40 and 50°C of orange juice with 10 ppm of dissolved oxygen.

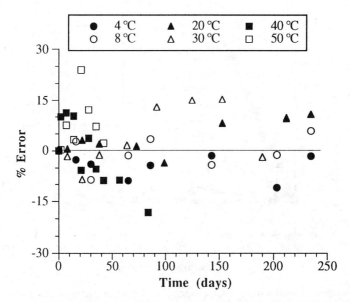

Figure 4. Errors in the prediction of the ascorbic acid values during storage at 4, 8, 20, 30, 40 and 50°C for an orange juice with 1 ppm of dissolved oxygen.

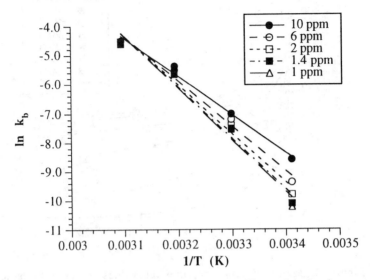

Figure 5. Arrhenius plot for the rate constant of browning development $(k_b)$

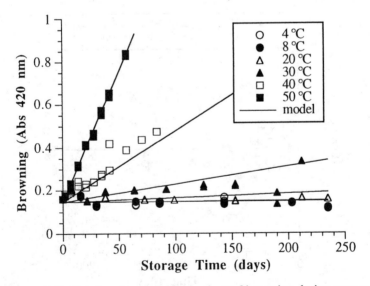

Figure 6. Predicted and experimental values of browning during storage at 4, 8, 20, 30, 40 and 50°C of orange juice with 6 ppm of dissolved oxygen.

(*4, 9, 10*) and it is probably one of the first significant off-flavours to be noticed during storage of orange juice. The equation (3) showed to be adequate to describe the change in PVG concentration for all storage temperatures.

$$P= P_\infty \, (1-\exp(-k_p t^2))$$ (3)

In this equation,
P (ppb) represents the concentration of PVG at time t (days),
$P_\infty$ (ppb) the maximum concentration of PVG and

$K_p$ - rate constant (days$^{-1}$)

The maximum value of PVG was only reached in the experiments at 50°C and its value (110 ppb) was used for all the other experiments.
The variation of the rate constant for PVG with temperature showed clearly two distinct Arrhenius profiles, one between 20 and 30°C and other between 30 and 50°C (Figure 7). The parameters of the Arrhenius equation did not vary with the oxygen initially dissolved in the juice, showing that the initial amount of dissolved oxygen in the juice does not influence the off-flavour development. Kacem and coworkers (*11*) verified that the deaeration of orange juice results in slightly increased retention of ascorbic acid, but had no effect on sensorial quality of the juices. An activation energy value of 83 kJ/mole was obtained for the low temperature range. The difference in the activation energies at high and low temperature makes a big difference in the formation of PVG at high and low storage temperatures as we can see in Figure 8. These plots also show the adequacy of the model to describe the experimental values.

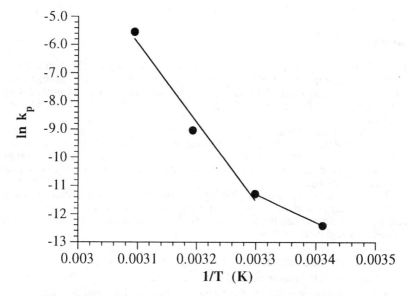

Figure 7. Arrhenius plot for the rate constant of PVG formation ($k_p$)

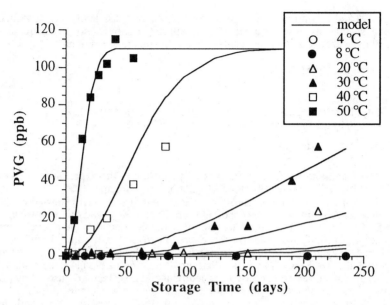

Figure 8. Predicted and experimental values of PVG concentration during storage at 4, 8, 20, 30, 40 and 50°C.

## Conclusions

The models developed are adequate to predict the ascorbic acid degradation, browning development and PVG formation in aseptically packed orange juice with initial concentration of dissolved oxygen between 1 and 10 ppm during storage from 4 to 50°C. Errors in the predictions lower than 25% were obtained for most of the cases.

The models developed can be used together, and whenever the endpoints that define shelf-life for the ascorbic acid, browning and PVG are defined, a computer programme may allow one to determine for how long aseptically packed orange juice can be stored, as well as to predict the quality degradation during storage at different temperatures.

The mathematical procedure developed in this work can be further extended to other packages and other processing conditions and a general predictive computer programme can be developed. Such a programme is a useful tool for food industry and consumers to:

- evaluate the effect of adding new additives on product,
- to identify the requirements for the package,
- to determine the 'Best Before' data for consumer information, and
- to ensure that consumed product complies with nutritional labelling.

## Literature Cited

1.  Fu, B.; Labuza, T.D. *Food Control*. **1993**, *4*, 125-133
2.  Singh, R.P.; Heldman, D.R. *Trans. of ASAE*. **1976**, *19*, 178-184
3.  Meydav, S.; Saguy, I.; Kopelman, I. J. *J. Agric. Food Chem.* **1977**, *25*, 602-604
4.  Lee, H. S.; Nagy, S. *J. Food Sci.* **1990**, *55*, 162-166
5.  Labuza, T.D.; Riboh, D. *Food Techn*. **1982**, *36*, 66-74
6.  Nagy, S.; Smoot, J. *J. Agric. Food Chem.* **1977**, *25*, 135-138
7.  Tannenbaum. In *Principles of Food Chemistry*; Fennema, O., Ed.; Marcel Dekker, Inc: New York and Basel, 1975, vol.1, pp 357-360
8.  Tatum, J. H.; Nagy, S.; Berry, R. E. *J. Food Sci.* **1975**, *40*, 707-709
9.  Naim, M.; Streim, B. J.; Kanner, J.; Peleg, H. *J. Food Sci.* **1988**, *53* , 500-503(512)
10.  Peleg, H.; Naim, M.; Zehavi, U.; Rouseff, R. L.; Nagy, S. *J. Agric. Food Chem* **1992**, *40*, 764-767
11.  Kacem, B.; Matthews, R. F.; Crandall, P. G.; and Cornell, J. A. *J. Food Sci.* **1987**, *52*, 1665-1667 (1672)

# Chapter 11

# Characteristic Stale Flavor Formed While Storing Beer

Fumitaka Hayase[1], Koichi Harayama[2], and Hiromichi Kato[3]

[1]Department of Agricultural Chemistry, Meiji University,
1-1-1 Higashi-mita, Tama-ku, Kawasaki, Kanagawa 214, Japan
[2]Nagoya Brewery, Asahi Breweries Ltd., Moriyama-ku, Nagoya 463, Japan
[3]Department of Food Science, Ohtsuma University, Chiyoda-ku,
Tokyo 102, Japan

When beer was stored for 0-4 weeks at 37°C, a stale flavor was generated with the increasing storage period. 132 peaks detected by a GC analysis were used as variables for a multivariate analysis. 15 compounds, including 2-furfuryl ethyl ether (2-FEE) and (E)-2-nonenal (E2N) were extracted as components that were highly correlated with the formation of the stale flavor. The results of the sensory evaluation revealed that the stale flavor was closely reproduced when 2-FEE and E2N were added together to beer. The experimental results on adding furfuryl acetate (FAC) and furfuryl alcohol (FAL) to beer indicate that 2-FEE was generated by the reaction between ethanol and FAL, which was formed by the hydrolysis of FAC, during the storage of beer.

Flavor plays a very important role in the quality of beer, and particularly off-flavors are a great problem in beer. Recently, Kamimura and Kaneda (1) have systematically reviewed on off-flavors in beer. Off-flavors that are produced during storage are expressed as a stale flavor. Drost et al. (2) and Clapperton (3) have attempted to express the stale beer flavor by using terms such as cardboard, metalic, sulfury, astringent, aldehydic, and musty. Cardboard flavor has been considered to be formed not only by (E)-2-nonenal, but also by other carbonyls such as (E)-2-heptenal (4). Therefore many compounds related to the stale flavor are considered to be generated in beer during storage. In a previous paper (5), GC headspace analysis of beer showed that nine compounds including furfural were extracted as the volatile components which take part in the cardboard flavor by multivariate analysis. However, (E)-2-nonenal and other unknown compounds were not identified in the headspace. This present paper reports the identification and the formation mechanism of some important compounds that were highly correlated with the stale flavor by a multivariate analysis by using a new trapping method for medium- and high-boiling-point volatiles in beer.

## Experimental

**Preparation of the beer sample.** Bottled domestic draft beer samples in five groups for different periods were used. Samples which were stored at 0° C were used as controls, and the others were stored at 37° C for 1, 2, 3, and 4 weeks.

**Sensory test.** The samples were evaluated for the intensity of stale flavor by an experienced panel after holding a sample in the mouth. The number of panelists was 10. The sum of the scores for each sample was divided by the number of panelists, and the resulting average is expressed as the sensory score. The difference in average scores between two neighboring groups was statistically analyzed by using the t-test (0.05).

**Sampling apparatus for volatile component.** Beer(600ml) placed in a Flash Evaporator (a thin film type, MF-80, Tokyo Rika Co. Ltd.,), was distilled under a reduced pressure of 5 mmHg for 1 h with 1 ml of silicon antifoam (TSA737K, Toshiba Silicon Co. Ltd.,). The volatiles were collected in a trap at -30℃ for 1h, a distillate of approximately 500 ml being obtained. After saturating with sodium chloride, liquid-liquid extraction was carried out for 20 min with 50 ml $CCl_3F$, which had been purified by distillation in advance. While organic layer was directly vaporized in a water bath at 40℃ for 1h under nitrogen at 100 ml/min, the vaporized volatiles were collected in a trap packed with Tenax TA (GL Science) according to the headspace trapping method described in a previous paper *(5,6)*.

## Capillary gas chromatography (GC) and gas chromatography-mass spectro-metory (GC-MS).

**GC.** A Shimadzu GC-8A gas chromatograph with a flame ionization detector. The GC conditions were as follows: column: 50m x 0.25mm i.d. fused silica capillary column with chemically bonded Carbowax 20M (GL Science Inc.), carrier gas: N2, flow rate, 1.56 ml/min and split ratio: 40.6:1. The oven temperature was held at 50° C for 5 min and then programmed to 190° C at 2° C/min.

The injection port and detector temperatures were kept at 200° C.

**GC-MS.** JEOL-DX303 apparatus (Japan Electron Optic Laboratory) was used, the carrier gas being helium. The other GC conditions were the same as those just described, and an MS analysis of electron impact (EI) at 70eV was conducted.

**Multivariate analysis.** The multivariate analysis was run with the BMDP programs by Dixon *(7)* on a HITAC M-200H/VOS3 system.

   **Factor analysis (FA).** FA was run with BMDP4M program *(7)* using varimax rotation *(8)* of log-transformed GC data.

   **Stepwise discriminant analysis (SDA).** SDA was run with the BMDP7M program *(7)*. The F-value for selecting a peak to enter in the classification functions or to remove from already classified functions was set as 4.0. The maximum step number was 10, and the precision of classification at every step was tested by the Jackknifed classification method *(8)*.

**Results and discussion**

**Sensory evaluation.** The results of sensory tests showed that intensity of the stale flavor in beer increased with storage period, and five groups (A: without storage; B: stored at 37° C for 1 week; C: stored at 37° C for 2 weeks; D: stored st 37° C for 3 weeks; E: stored at 37° C for 4 weeks) were analyzed by t-tests at level of significance of 0.05 or less.

**GC analysis.** It is possible to perceive a stale flavor in stored beer that not only involves the cardboard flavor that was studied with a headspace analysis in a previous paper *(5)* but also an astringent, aldehydic and musty flavor in the odor concentrate. Figure 1 shows typical gas chromatograms of beer with and without storage at 37° C. The 132 peaks detected by GC were used as variables for the subsequent multivariate analysis *(9)*. Table I shows the compounds identified in this study as 14 alcohols, 11 esters, 6 aldehydes, 4 ketones, 2-furfuryl ethyl ether, 2-acetylfuran, γ-nonalactone, and 4 acids.

**Factor analysis (FA).** A multivariate analysis of beer has been applied to recognize different brands of beer *(10)*. FA was conducted on the basis of log-transformed GC peak area data with varimax rotation. Eigen values surpassed 7.0 up to the fifth factor, the cumulative proportion reaching 75% on the basis of 5 extracted factors. This indicates that 132 peaks in the original GC data could be grouped as 5 factors with 25% loss of information.

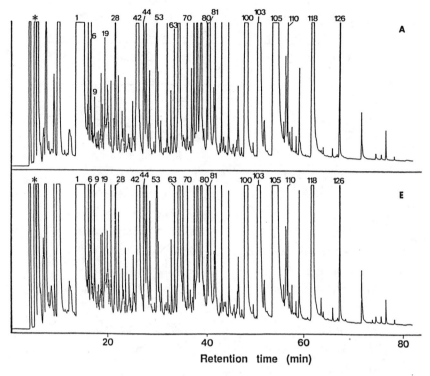

**Figure 1.** Gas chromatograms of volatiles in beer obtained using flash evaporator

(A): without storage,

(E): stored at 37°C for 4weeks.

* : ethanol

Each sample was plotted on the basis of the second and third factors as shown in Fig. 2. Each sample shifted with storage period from the negative to positive direction on the second factor axis. This suggests that the second factor axis indicates the storage time of beer. The rotated factor loadings for each peak were plotted for the second and third factor axis as shown in Fig. 3. Fifteen compounds that had rotated factor loadings of more than 0.5 on the second factor axis were extracted as shown with a circle in Fig. 3. These are considered to be highly correlated with the formation of the stale flavor. On the other hand, three compounds that had rotated factor loadings of less than -0.5 were also extracted. Peak Nos. 6, 9, 16, 26, 38, 44, 53, 59, 60, 63, 64, 67, and 114 were respectively

Table I.  Volatile Components Identified in Beer by
Using a Flash Evaporator

| Peak no. | Compound | Peak no. | Compound |
|---|---|---|---|
| Acids | | Esters | |
| 78 | Butyric acid | 8 | Hexyl acetate |
| 81 | Isovaleric acid | 19 | Methyl lactate |
| 103 | Caproic acid | 26 | Ethyl lactate |
| 118 | Octanoic acid | 28 | Ethyl heptanoate |
| | | 29 | Heptyl acetate |
| Alcohols | | 42 | Ethyl caproate |
| * | Ethanol | 46 | Isoamylcaproate |
| 1 | Isoamyl alcohol | 60 | Furfuryl acetate |
| 4 | Pentanol | 79 | Ethyl decanoate |
| 16 | 3-Methyl-2-buten-1-ol | 100 | 2-Phenylethyl acetate |
| 27 | Hexanol | 126 | 2-Phenylethyl hexanoate |
| 35 | 3-Methyl hexanol | | |
| 47 | Heptanol | | |
| 55 | 2-Ethyl-1-hexanol | Ether | |
| 63 | Linalool | 9 | 2-Furfuryl ethyl ether |
| 66 | Octanol | | |
| 70 | 4-Terpineol | Furan | |
| 80 | Furfuryl alcohol | 53 | 2-Acetylfuran |
| 105 | 2-Phenylethyl alcohol | Ketones | |
| 110 | Dodecanol | 6 | 2-Methyltetra-hydrofuran-3-one |
| Aldehydes | | 22 | 1-Methyl heptenone |
| 38[a] | (E)-2-Octenal | 24 | 2-Methyl-2-hepten-6-one |
| 44 | Furfural | | |
| 57 | Benzaldehyde | 32 | 2-Nonanone |
| 59 | (E)-2-Nonenal | | |
| 64 | 5-Methylfurfural | Lactone | |
| 67[a] | (E,Z)-2,6-Nonadienal | 114 | $\gamma$-Nonalactone |

a: Tentatively identified compounds     *  Indicated in figure 1

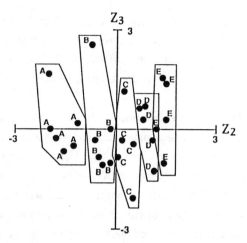

Figure 2. Distribution of 27 samples on the basis of the second factor (Z2) and third factor (Z3) from factor analysis with varimax rotation of log-transformed peak area data.
A: without storage, B: stored at 37° C for 1 week, C: stored at 37° C for 2 weeks, D: stored at 37° C for 3 weeks, E: stored at 37° C for 4 weeks.

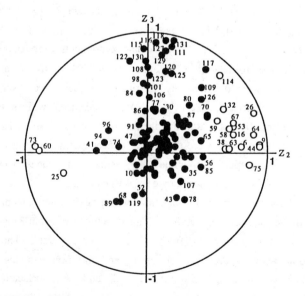

Figure 3. Relationships between rotated factor loadings for second and third factors.
The number represents peak number of GC.
○ : Factor Loading is not less than ± 0.5 ;
● : Factor Loading is less than ± 0.5 .

identified    as    2-methyltetrahydrofuran-3-one, 2-furfuryl ethyl ether,   3-methyl-2-
buten-1-ol, ethyl lactate,   (E)-2-octenal,   furfural,   2-acetylfuran,   (E)-2-nonenal,
furfuryl   acetate,   linallol,   5-methylfurfural,   (E,Z)-2,6-nonadienal, and $\gamma$ -
nonalactone, which had a markedly high rotated factor loading.

**The stepwise discriminant analysis.** The stepwise discriminant analysis was
run on the basis of eighteen compounds which had a  rotated   factor loading of more
than 0.5 and less than -0.5 on the second factor   axis.   Furfuryl acetate, 5-
methylfurfural, and unknown compound (Peak No.    73) were selected by the
stepwise discriminant anlysis. At step 1, furfuryl acetate, with the largest F-value, was
entered into the equation as the most discriminative compound. The results of the jackknifed   classifi-cation at step 3 showed that the samples could be discriminated in each
group with 100% correctness on  the  basis  of  the  three  components. Furfuryl
acetate had  high  F-value,  and  was  extracted  as   an   important  index  for
discriminating beer based on storage period.

**Formation mechanism for 2-furfuryl ethyl ether.** In order to investigate the
formation mechanism for 2-furfuryl ethyl ether (2-FEE), 100   $\mu$ l  of furfuryl acetate
(FAC) or 10  $\mu$ l of furfuryl alcohol (FAL) was added   to a 0.2M McIlvaine
phosphate buffer at pH 4.2 in the presence of  ethanol or to fresh beer, and the
samples were stored   at   37° C.   Ten   volatile components were found to be closely
correlated with  the  formation  of the stale  flavor  by  a  factor  analysis,  and
the  components   were  quantitatively determined in each · beer   sample   during
storage.   Each component, except for FAC,   increased  with  storage,  furfural
increasing to over 1,000 ppb after storage for  4  weeks.  Ethyl  lactate  and  2-
methyltetrahydrofuran-3-one were detected at concentrations   of   appro-ximate 500
ppb and 150 ppb, respectively,  while   the   other   components   were detected at
concentrations of 0.3-50 ppb after storage for 4 weeks *(11)*.

Table II shows a sensory evaluation after adding each component to fresh beer.
No change in the  aroma  after  adding  2-methyltetrahydro-furan-3-one, 3-methyl-2-
buten-1-ol or 2-acetylfuran to fresh   beer   was   apparent. The   aroma   was
distinguishable between the   control   beer   and the beer samples with   ethyl
lactate,   furfural,   linalool,   5-methyl-furfural,  and  $\gamma$ -nonalactone respectively
added, but no stale flavor was detected in each beer sample. An astringent note, this
being   part  of the stale flavor, appeared after adding 2-furfuryl ethyl ether   (2-FEE)
to the control beer. 2-FEE is considered to have  been  partly  respon-sible for the
stale flavor, although it was not   completely   reproduced by the addition of 2-FEE

Table II.  Sensory Evaluation after Adding Selected
Compounds to Fresh Beer

| Sample No. | Compound | Added quantity (ppb) | Sensory |
|---|---|---|---|
| 1 | 2-Methyltetrahydrofuran-3-one | 150.0 | N.D. |
| 2 | 3-Methyl-2-buten-1-ol | 15.0 | N.D. |
| 3 | Ethyl lactate | 400.0 | Stale hop aroma |
| 4 | Furfural | 1000.0 | Sweetish |
| 5 | 2-Acetylfuran | 20.0 | N.D. |
| 6 | Linalool | 2.5 | Powdery |
| 7 | 5-Methylfurfural | 2.0 | Musty |
| 8 | $\gamma$-Nonalactone | 14.0 | Papery |
| 1-8 | 8 Compounds combined | 1603.5 | Weakly stale |
| 9 | 2-Furfuryl ethyl ether(2-FEE) | 6.0 | Astringent, stale |
| 10 | (E)-2-nonenal  (E2N) | 0.3 | Cardboard, stale |
| 9+10 | 2-FEE + E2N combined | 6.3 | Strongly stale |

N.D.: not distinguishable

alone. The sensory test  for  2-FEE  indicated that the threshold value for 2-FEE in beer was approximately  2.5  ppb, and the astringent stale flavor was detected at a concentration of over 6 ppb of 2-FEE.

(E)-2-nonenal (E2N) is known as a beer volatile component having a cardboard flavor (*2, 12-14*). In this study, the stale  flavor  appeared more strongly by the addition of E2N than by 2-FEE. However, the  stale flavor could not be completely reproduced by  adding  E2N  alone.   The addition of  both  E2N and 2-FEE intensified the stale flavor, which  was stronger than that by the addition of E2N or 2-FEE alone.

Figure 4 shows the  amounts  of  furfuryl  acetate  (FAC)  and  2-furfuryl ethyl ether (2-FEE) in the beer samples with  or  without  the addition of 0.5 ml of FAC. FAC remained at a concentration of 47 ppb or 4 ppb in the beer samples before storage and those stored at  25° C  for 70  days  without  the  addition  of  FAC, 2-FEE  being  formed  at  a concentration of 2 ppb in the former case, and at 11 ppb in the  latter case. However, 2-FEE was formed at a concentration of 0.8  ppb  in

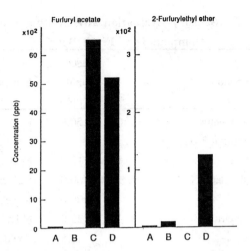

Figure 4.  Changes  in  the  amounts  of  furfuryl  acetate  and
2-furfuryl  ethyl  ether  by   the   addition   of  furfuryl
acetate (FAC) to fresh beer.
A and B: beer without FAC
C and D: beer with FAC
A and C: beer without storage
B and D: beer stored at 25° C for 70 days

the former case and 125  ppb  in  the  latter  case  when  FAC  was  added.
Therefore, 2-FEE increased by 9 ppb in the beer without the addition of FAC   and
by  124.2  ppb  in  the  beer  with  the  addition  of   FAC.   Accordingly, 2-
FEE is considered to have been  formed  through  FAC  in beer.

2-FEE was also markedly increased   with storage after the addition of 0.1 ml of
furfuryl alcohol (FAL) to beer as shown in Fig. 5.

These results indicate that 2-FEE was formed by  the   reaction of ethanol and
FAL, which itself was   formed  by  the  hydrolysis  of  FAC during the   storage
of beer as shown in Fig. 6.   This   scheme  is  also supported by the experimental
results on  adding  FAC  and  FAL   to  a phosphate buffer in the presence of 5%
ethanol (11). FAL might be   also have been formed by the Maillard reaction (15).
FAC is  thought  to  be produced by FAL and  acetic  acid  during  the
fermentation  of  beer, because of the formation of various volatile acetate compounds
in beer. Another possibility that the direct interaction  of  FAC  with   ethanol leads to
the formation of 2-FEE does not rule out.

Figure 5. Changes in the amounts of 2-furfuryl ethyl ether by the addition of furfuryl alcohol to fresh beer.
A: beer stored at 37° C for 6 days without furfuryl alcohol
B: beer with furfuryl alcohol without storage
C: beer stored at 37° C for 6 days with furfuryl alcohol

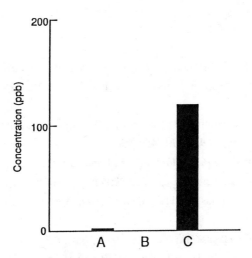

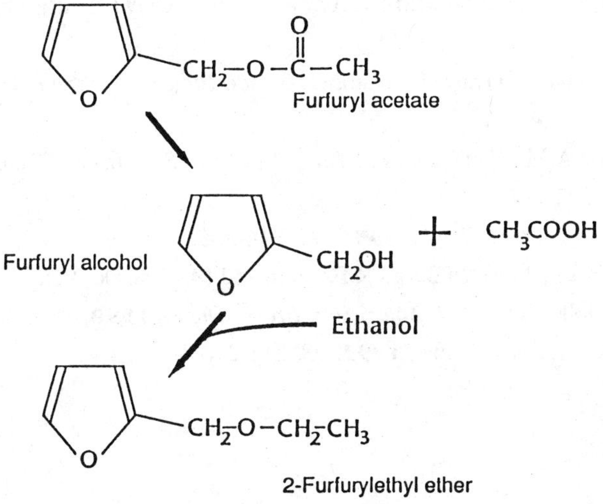

Figure 6. Proposed Mechanism for the Formation of 2-Furfuryl Ethyl Ether in Beer during Storage

## Literature cited

1. Kamimura, M., Kaneda, H. *In Off Flavors in Foods and Beverages:* Charalambous, G., Ed., Elsevier, Dev. Food Sci., **1992**, *28*, 433-472.

2. Drost, B.W.; Van Eerde, P.; Hoekstra, S.F.; Strating, J. *Proc. Eur. Brew. Conv.* **1971**, *13*, 451-458.

3. Clapperton J. F. *In Progr. Flavor Res.*, Land, D.G.; Nursten, H.E. Eds., Applied Science Pub. London, **1979**, pp 1-14.

4. Murakami, A.; Goldstein, H.; Chicoye, E. *J. Am. Soc. Brew. Chem.* **1986**, *44*, 33-37.

5. Harayama, K.; Hayase, F.; Kato, H. *Agric. Biol. Chem.* **1991**, *55*, 393-398.

6. Harayama, K.; Hayase, F.; Kato, H. *Biosci. Biotech. Biochem.* **1994**, *58*, 2246-2247.

7. In BMDP Biomedical Computer Programs; Dixon, W.J. Ed., Health Sciences Computing Facility, University of California Press, Los Angeles, **1975**.

8. Kaiser, H.F. *Educat. Psych. Meas.* **1959**, *19*, 413-420.

9. Harayama, K.; Hayase, F.; Kato, H. *Biosci. Biotech. Biochem.* **1994**, *58*, 1595-1598.

10. Siebert, K.J.; Stenroos, L.E. *J. Am. Soc. Brew. Chem.* **1989**, *47*, 93-101.

11. Harayama, K.; Hayase, F.; Kato, H. *Biosci. Biotech. Biochem.* **1995**, *59*, 59, 1144-1146.

12. Jamieson A.M.; Van Gheluwe J.E.A. *Proc. Am. Soc. Brew. Chem.*, **1970**, 192-197.

13. Narziss, L. *J. Inst. Brew.*, **1986**, *92*, 346-353.

14. Drost, B.W.; Van Den Berg, R.F.; Freijee, J.M.; Van Der Velde, E.G.; Hollemans, M. *J. Amer. Soc. Brew. Chem.*, **1990**, *48*, 124-131.

15. Moir, M. *J. Inst. Brew.*, **1992**, *98*, 215-220.

Chapter 12

# Use of Volatile Aldehydes for the Measurement of Oxidation in Foods

Jon W. Wong and Takayuki Shibamoto

Department of Environmental Toxicology, University of California,
Davis, CA 95616

Volatile aldehydes are formed as secondary by-products of lipid peroxidation or products of other oxidation processes. Due to their effects on the flavor quality in foods and their toxicological and biological significance, there is a need to develop noninvasive methods for assessing these products. Two analytical methods have been developed which utilize derivatization of the aldehydes and analysis by capillary gas chromatography with nitrogen-phosphorus detection. N-Methylhydrazine and cysteamine derivatization with malondialdehyde and straight chain aldehydes, respectively, and subsequent GC analysis of the corresponding derivatives have been used to study a variety of systems such as liquids (alcoholic drinks), food stuffs (oils and fats), antioxidant-treated model systems, and LDL (low density lipoprotein) oxidation. These methods allow one to characterize specific aldehyde products and their interactions and determine the mechanisms of formation in complex biological and food systems.

Volatile aldehydes formed by oxidation processes are important in the taste, flavor, and aroma of foods. In addition to their importance in food quality, low-molecular weight aldehydes such as formaldehyde, acetaldehyde, and malonaldehyde are also known to play important biological and physiological roles. Aldehydes are strong electrophiles that are capable of rapid reactions with nucleophiles such as nucleic acids, amino acids, and proteins. In food systems, carbonyls such as aldehydes and ketones produce heterocyclic flavor compounds with amino acids and proteins via Maillard-type reactions in cooked foods. Furthermore, in biological systems and in some food systems, lipid peroxidation has been implicated in DNA and protein modification, radiation damage, aging, modification of membrane structure, tumor initiation, and the deposition of arterial plaque associated with low-density lipoprotein (LDL) modification (1).

Lipid peroxidation of the membranes and fats plays an important role in the production of these aldehydes in cooked foods. Complex chemical and physical processes such as cooking can cause fat and lipid deterioration in which the nutritional and toxicological values of the food may be affected. Formaldehyde is

0097–6156/96/0631–0129$15.00/0

widely found in both foods and in the environment and has also been shown in animal studies to be a carcinogen and a skin and pulmonary irritant *(2)*. Acrolein has been found in various foods such as molasses, salted pork, mackerel, and white bread and is a known mutagen *(3)*. Malondialdehyde, a carcinogen, mutagen, and well-known product of lipid peroxidation and prostagladin biosynthesis has been reported in numerous lipid- and fat-rich foods.

There is a major effort to analyze volatile aldehydes because there have been some difficulties in the determination of their levels in foods, as well as in biological systems and in the environment. This chapter will examine the relevance of assessing oxidative damage using analytical techniques used in identifying and quantitating low molecular weight aldehydes in a variety of model, food, and biological systems.

## Formation of Volatile Aldehydes

Autoxidation is the most recognized process involved in the oxidative deterioration of lipids. This process consists of chain initiation, propagation, and termination steps, as shown below:

$$
\begin{array}{ll}
\text{Initiation:} & \text{LH} + \cdot\text{OH} \longrightarrow \text{L}\cdot + \text{H}_2\text{O} \\
\text{Propagation:} & \text{L}\cdot + \text{O}_2 \longrightarrow \text{LOO}\cdot \\
 & \text{LOO}\cdot + \text{LH} \longrightarrow \text{L}\cdot + \text{LOOH} \\
\text{Termination:} & 2\,\text{LOO}\cdot \longrightarrow \text{nonradical products}
\end{array}
$$

The initiation step usually involves the abstraction of the hydrogen atom by the free radical, such as the hydroxyl radical ($\cdot$OH), adjacent to a double bond of a polyunsaturated fatty acid (LH) to form a lipid free radical, L$\cdot$. Once the hydrogen atom is abstracted, propagation involves the reaction of the lipid free radical with molecular oxygen to form the hydroperoxy lipid radical (LOO$\cdot$), which in turn can remove hydrogen from another lipid molecule. Termination occurs when the radicals react to form nonpropagating intermediates and relatively stable nonradical products. The major primary products are the lipid hydroperoxides, which can undergo rearrangement and cyclization to form hydroperoxyepidioxides and biscycloendoperoxides *(4)*. These primary products are relatively unstable and can break down to form a variety of secondary breakdown products such as carbonyls, saturated and unsaturated acids and esters, hydroxy acids, alkanes, and alcohols *(5)*. The mechanism for the deterioration of polyunsaturated fatty acids to form malondialdehyde, an important marker for lipid peroxidation, has been postulated to involve the formation of these intermediate bicycloendoperoxides, which subsequently break down to small molecular weight products such as dicarbonyls *(6)*.

## Analysis of Volatile Aldehydes

Early methods to analyze aldehydes involved spectrophotometric assays, which are advantageous in their simplicity and and convenience, but are not specific and not as sensitive for determining low levels of aldehydes. The most commonly used analytical method, the thiobarbituric acid (TBA) assay, is based on the reaction of carbonyls with TBA to form a chromagen that can be measured spectrophotometrically at 500 - 600 nm. Although the TBA assay was used to measure malondialdehyde, other aldehydes can interfere with the method by reacting with TBA to form the same pink chromagen. Other non-carbonyl compounds such as lipids, carbohydrates, proteins, pigments, and metal ions can also interfere with the assay *(7)*. In addition, the TBA assay requires highly acidic and elevated temperature conditions, resulting in possible alterations of the compound of interest and increased aldehyde formation. Other methods were used to determine carbonyls with or without derivatization with TBA. To improve specificity, various studies have utilized milder conditions and separation methods such as solid-phase extraction, gas chromatography, and high-performance liquid chromatography (HPLC), to remove interfering compounds and eliminate excessive compounds formed as a result of the conditions employed (*8, 9*).

Combined derivatization and chromatographic methods have been utilized for analysis of carbonyls. The most common agent used to analyze aldehydes is 2,4-dinitophenylhydrazine derivatization, followed by separation by HPLC with UV or fluorescence detection or gas chromatography with nitrogen-specific detection (*10, 11*). However, there are some disadvantages to this method. Derivatization requires strongly acidic conditions that may cause undesirable side reactions, such as the decomposition of carbohydrates or proteins in the cases of food samples. GC methods require on-line injection or negative ion mass spectrometry due to the thermally unstable derivatives (*11, 12*), while LC analysis is limited due to a lack of detection systems.

Recently, aldehydes have been derivatized to thiazolidine derivatives with cysteamine and then analyzed by gas chromatography (GC) with a fused capillary column and a nitrogen-phosphorus specific detector (NPD) (*3*). This method has several advantages to other derivatization methods: 1) only one derivative is formed from an individual aldehyde; 2) derivatization occurs rapidly under mild conditions; 3) derivatives can be separated by GC; and 4) the excess cysteamine reagent does not interfere with the chromatography (*13*). The physical properties and NPD detectability of these thiazolidine derivatives show that these compounds are sufficiently stable for GC analysis and that detection is roughly proportional to molecular weight, as the relative nitrogen content in the molecule decreases with increasing molecular weight (*13*).

A second method which is used to analyze volatile α,β-unsaturated aldehydes such as acrolein and 4-hydroxynonenal and dicarbonyls such as malondialdehyde, utilizes *N*-methylhydrazine derivatization (Figure 1) and GC-NPD analysis (*14*). This method was originally used to measure free malondialdehyde during UV irradiation of hexane solutions containing free fatty acids, corn oil, or beef fat, but has been used successfully for analysis of malondialdehyde in biological and food systems (*15–17*).

## Studies of Volatile Aldehydes in Model and Food Systems

Cysteamine and *N*-methylhydrazine derivatization with GC-NPD analysis were used to analyze volatile aldehydes in model, food, and biological systems. A model system consisting of arachidonic acid and a Fenton oxidation system ($FeCl_2$, $H_2O_2$) was analyzed using NMH derivatization, continuous liquid-liquid extraction, and GC-NPD. Both MDA and 4-hydroxynonenal, a common hydroxyalkenal product with carcinogenic and cytotoxic effects which formed in oxidized lipids, were present in the extract as the derivatives, 1-methypyrazole and 5(1'-hydroxyhexyl)-1-methyl-2-pyrazoline, respectively (*18*). Since it was shown to be a sensitive assay for quantitating MDA specifically, the method was found to be an effective alternative to the commonly used TBA assay. Another application of this method was used to assess the effects of antioxidants using MDA as a marker. Both the known antioxidant, α-tocopherol, and a test agent derived from green barley leaves, 2-O"-glycosylisovitexin (2-O"-GIV) were tested in a phospholipid model system or fish oil with a Fenton oxidation system (*19*). The results showed that 1) both a-tocopherol and 2-"-GIV were effective antioxidants in inhibiting lipid peroxidation; 2) the antioxidant activities of both compounds were concentration-dependent;and 3) 2-O"-GIV was consistently shown to be a more effective antioxidant than α-tocopherol in the cod liver system.

Other potential markers for lipid peroxidation include straight-chain, low-molecular weight aldehydes, such as formaldehyde, acetaldehyde, and propanal, and α,β-unsaturated carbonyls, such as acrolein. Acrolein and other α,β-unsaturated carbonyls can be derivatized to form N-methylpyrazolines, while the aldehydes are converted to

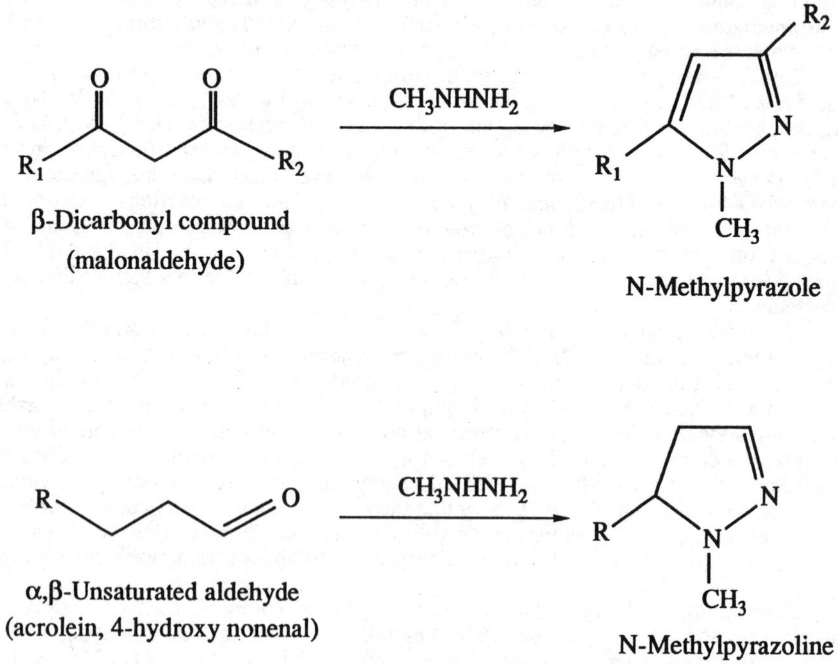

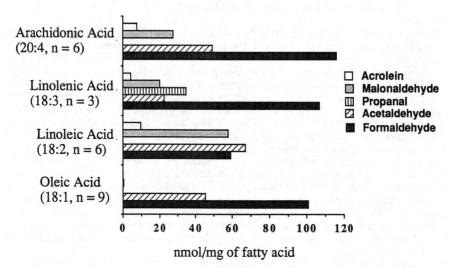

β-Dicarbonyl compound
(malonaldehyde)

N-Methylpyrazole

α,β-Unsaturated aldehyde
(acrolein, 4-hydroxy nonenal)

N-Methylpyrazoline

Figure 1. Stable derivatives from carbonyl compounds with N-methylhydrazine.

Figure 2. Amount of volatile aldehydes formed from fatty acids by the Fenton's reagent.

alkylthiazolidine derivatives with cysteamine. From Figure 2, a variety of volatile aldehydes are formed in various fatty acids oxidized with Fenton's reagent. The results reveal that no major trends were observed, although formaldehyde was formed in the largest amounts, and that acrolein, malondialdehyde, and propanal do not form appreciably in the oleic acid system. UV-irradiation of various lipids follow the general trend that the higher degree of unsaturation of the lipid leads to an increase of individual aldehydes and total aldehyde content (Figure 3).

During heat treatment of foods, various aldehydes are formed in the head space. When pork fat was heated, various aldehydes such as formaldehyde, acetaldehyde, propanal, butanal, pentanal, and hexanal, were identified and quantitated (20). Hexanal was found in the largest amount, followed by pentanal and the other lower chain aldehydes. Both hexanal and pentanal have been commonly used as markers of lipid peroxidation in other food systems (21). However, irradiated cod liver oil produced malondialdehyde of which level was nearly five times higher than that of other aldehydes (22), suggesting that different oxidative mechanisms are involved and that more than one specific marker may be required to understand the formation of various aldehyde species.

**Other Oxidation Systems**

Although aldehydes are common secondary products of lipid peroxidation, there are cases in which the aldehyde formation may be a result of some other oxidative mechanism. Cysteamine derivatization of acetaldehyde to the derivative 2-methylthiazolidine followed by extraction and GC-NPD analysis was used to measure significant levels of acetaldehyde in alcoholic beverages (23). Due to the negligible presence of lipids in alcoholic beverages, the formation of acetaldehyde was postulated to be a byproduct of ethanol oxidation, rather than lipid oxidation. Due to the characteristics of these beverages, it is conceivable that ethanol, under oxidative decomposition from hydroxyl radical, can be easily converted to acetaldehyde, as shown in Figure 4. Currently, acetaldehyde in alcoholic beverages over prolonged time periods is being measure to correlate formation with oxidative deterioration and quality.

Derivatization/GC analysis of acetaldehyde has also been shown as an effective marker in biologically relevant oxidation systems. An oxidation system containing low density lipoprotein (LDL), Fenton's reagent, and ascorbic acid showed acetaldehyde levels nearly three times higher than in a system without ascorbic acid (24). This result suggest that the increase of acetaldehyde was due to the presence of ascorbic acid. Both Fenton's reagent and UV-irradiation of ascorbic acid were shown to induce the formation of acetaldehyde in a linear relationship. The mechanism of formation of acetaldehyde from L-ascorbic acid upon oxidation was proposed to involve decarboxylation to a L-xylosone intermediate, and subsequent dehydration to form the breakdown products 1,2-diketopropanoic acid and acetaldehyde. For the effect of UV irradiation, the formation of acetaldehyde involves the degradation of a 1,4 biradical formed by a Norrish type II photocleavage of a $\gamma$-hydrogen from L-xylosone. The results of this study showed the oxidative formation of acetaldehyde from ascorbic acid. Further studies show that LDL oxidation by Fenton's reagent could be inhibited by the antioxidants, ferulic acid and 2"-O-glycosylisovitexin, using acetaldehyde as a marker (Figure 5).

**Conclusions**

Derivatization and gas chromatographic methods were utilized to identify and quantitate low molecular weight volatile aldehydes in various model, food, and biological systems. Volatile aldehydes have been shown to be produced from the oxidation of lipid-rich foods, beverages, and lipoproteins. The methods employed in these studies

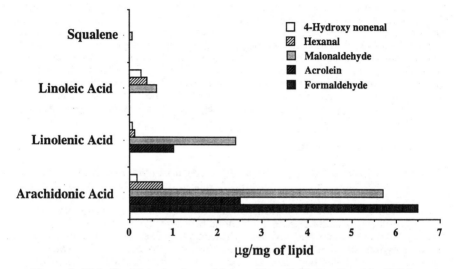

Figure 3. Volatile aldehydes formed from various lipids irradiated by UV ($\lambda$ = 350 nm).

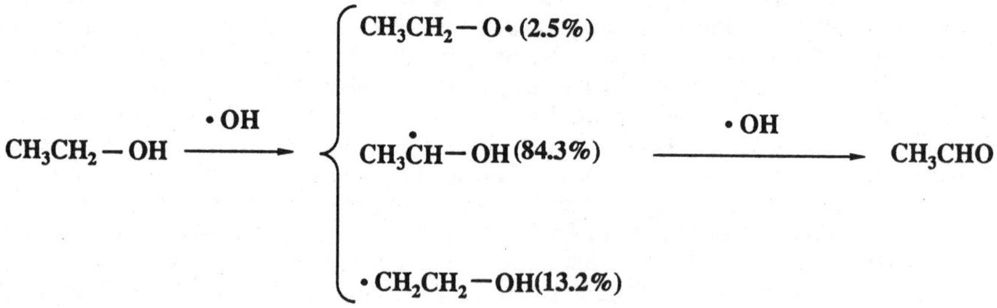

$$CH_3CH_2-OH \xrightarrow{\cdot OH} \begin{cases} CH_3CH_2-O\cdot (2.5\%) \\ \\ CH_3\dot{C}H-OH (84.3\%) \xrightarrow{\cdot OH} CH_3CHO \\ \\ \cdot CH_2CH_2-OH (13.2\%) \end{cases}$$

Figure 4. Formation of acetaldehyde from ethanol by a hydroxy radical.

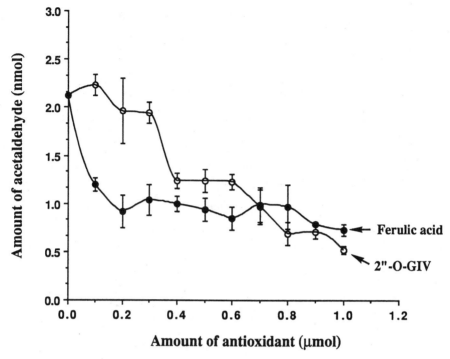

**Amount of antioxidant (μmol)**

Figure 5. Inhibition of acetaldehyde formation by antioxidants in low density lipoproteins oxidized with Fenton's reagent.

are useful to monitor the deterioration of various food products and to assess the inhibition activities of antioxidants.

## Literature Cited

1. Porter, N.A.; Caldwell, S. E.; Mills, K. A. *Lipids* **1995**, *30*, 277–290.
2. Heylin, M. *Chem. Eng. News* **1979**, *57*, 7.
3. Shibamoto, T. In: *Flavors and Off-Flavors, Proceedings of the 6th International Flavor Conference;* Charalambous, G., Ed.; Elsevier Science Publishers: Amsterdam, 1989, pp. 471-483.
4. Pryor, W. A.; Stanley, J. P.; Blair, E. *Lipids* **1976**, *11*, 370–379.
5. Frankel, E. N.; Neff, W. E.; Selke, E. *Lipids* **1981**, *16*, 279–285.
6. Dahle, L. K.; Hill, E. G.; Holman, R. T. *Arch. Biochem. Biophys.* **1962**, *98*, 253–261.
7. Hoyland, D. V.; Taylor, A. J. *Food Chem.* **1991**, *40*, 271–291.
8. Bird, R. P.; Draper, R. P. *Methods in Enzymology* **1984**, *105*, 299–305.
9. Raharjo, S.; Sofos, J. N.; Schmidt, G. R. *J. Food Sci.* **1993**, *58*, 921–924, 932.

10. Esterbauer, H.; Lang, J.; Zadravec, S.; Slater, T. F. *Methods in Enzymology* **1984**, *105*, 319–328.
11. Dalene, M.; Persson, P.; Skarping, G. *J. Chromatog.* **1992**, *626*, 284–288.
12. Thomas, M.J.; Robison, T. W.; Samuel, M.; Forman, H. J. *Free Radical Biol. Med.* **1995**, *18*, 553–557.
13. Ebeler, S. E.;Shibamoto T.; Osawa T. In: *Lipid Chromatographic Analysis*; Shibamoto, T., Ed.; Chromatographic Science Series 65; Marcel Dekker Inc.: New York, 1994, pp 223–249.
14. Umano, K.; Dennis, K. J.; Shibamoto, T. *Lipids* **1988**, *23*, 811–814.
15. Ichinose, T.; Miller, M. G.; Shibamoto, T. *Lipids* **1989**, *24*, 895–898.
16. Ebeler, S. E.; Hinrichs, S. H.; Clifford, A. J.; Shibamoto, T. *J. Chromatogr. B: Biomed. Appl.* **1994**, *654*, 9–18.
17. Wong, J. W.; Ebeler, S. E.; Isseroff, R. R.; Shibamoto, T. *Anal. Biochem.* **1994**, *220*, 73–84.
18. Tamura, H.; Shibamoto, T. *Lipids* **1991**, *26*, 170–173.
19. Nishiyama, T.; Hagiwara, Y.; Hagiwara, H.; Shibamoto, T. *J. Am. Oil Chem. Soc.* **1993**, *70*, 811–813.
20. Yasuhara, A.; Shibamoto, T. *J. Food Sci.* **1989**, *54*, 1471–1472, 1484.
21. Shahidi, F.; Yun, J.; Rubin, L. J.; Wood, D. F. *Can. Inst. Food Sci. Technol. J.* **1987**, *20*, 104–106.
22. Niyati-Shirkhodaee, F.; Shibamoto, T. *J. Am. Oil Chem. Soc.* **1992**, *69*, 1254–1256.
23. Miyake, T.; Shibamoto, T. *J. Agric. Food Chem.* **1993**, *41*, 1968–1970.
24. Miyake, T.; Shibamoto, T. *J. Agric. Food Chem.* **1995**, *43*, 1669–1672.

# Chapter 13

# Autoxidation of L-Ascorbic Acid and Its Significance in Food Processing

T. Kurata[1], N. Miyake[1], E. Suzuki[2], and Y. Otsuka[3]

[1]Institute of Environmental Science for Human Life, and [2]Department of Human Biological Studies, Ochanomizu University, 2–1–1 Otsuka, Bunkyo-ku, Tokyo 112, Japan
[3]Faculty of Education, Tottori University, Tottori, Japan

Oxidation rates of L-ascorbic acid (ASA) in a low dielectric solvent (MeOH) were determined, and initial autoxidation products were separated and identified. Mechanisms of ASA autoxidation were studied using the semi-empirical molecular orbital (MO) method. In this solvent, MeOH and in the absence of metals, ASA reacts with triplet oxygen yielding very similar to or nearly identical reaction products as those observed with singlet oxygen, suggesting the formation of ascorbate-2-peroxy-anion type intermediate.mechanisms of ASA autoxidation including the formation mechanism of superoxide anion radical were proposed, which were supported by the MO calculation.

ASA is an important antioxidant in food as well as in biological systems. In its chemical structure, ASA has an ene-diol group conjugated with a lactone carbonyl group, hence, it is a typical "aci-reductone" (1) that shows a strong reducing activity. Thus, ASA is susceptible to oxidation, and ASA present in various foods is usually very easily oxidized during processing or storage to give a mixture of complex reaction products, which strongly influence the quality of those foods. For instance, it is well known that ASA is involved in the discoloration of lemon, orange and other citrus juices and causes non-enzymatic browning reactions, and the presence of oxygen accelerates the oxidation of ASA, hence accelerates the deterioration of these fruit juices. Thus, the oxidation reaction, especially the autoxidation of ASA, i.e., the reaction of ASA with oxygen molecules, can be regarded as a marker reaction that may reflect the oxidative status of other constituent compounds in various foods.

In this paper, a preliminary study to evaluate oxidation rates of ASA in both aqueous and non-aqueous systems is presented. Reaction products of the initial autoxidation process of ASA were separated and identified. To elucidate the reaction mechanism

0097–6156/96/0631–0137$15.00/0

involved in the autoxidation process of ASA, and to obtain information on the reactivities and structures of rather unstable reaction intermediates produced in the autoxidation process, the semi-empirical molecular orbital (MO) method was employed. Some possible reaction mechanisms considered to be operative after the initial autoxidation process, involving dehydro-L-ascorbic acid (DHA), 2,3-diketo-L-gulonic acid (DKG) and 2-amino-2-deoxy-L-ascorbic acid (L-scorbamic acid; SCA) were also briefly discussed.

## Materials and Methods

ASA was obtained from a commercial source (Wako Pure Chemical Industries, Ltd) and further purified by recrystallization. Ultra-refined water with electrical resistance of 18 M$\Omega$ · cm was used throughout the experiment. Commercially obtained MeOH (Wako Pure Chemical Industries, Ltd, heavy metal contents: less than 0.5 ppb) was used, and when necessary, it was further purified by distillation.

**A preliminary study on autoxidation rates of ASA.** Very dilute ASA solutions (conc. 0.03 mg%) were prepared and incubated at 25, 30 and 35 ℃ for 30 min. The degradation rate of ASA was determined using HPLC with highly sensitive electrochemical detector (BAS LC-4B, 600 mV), and first order rate constants for the autoxidation of ASA were estimated.

**Separation and identification of initial autoxidation products of ASA in MeOH.** Recrystallized ASA was dissolved in 200 ml of MeOH at a concentration of 50 $\mu$M and reacted with oxygen by bubbling oxygen gas through the solution. The solution was kept at 25℃ throughout the autoxidation reaction. Oxygen gas was bubbled through the solution at a flow rate of 200 ml/min for 30 to 60 min, while the reaction chamber was kept in darkness. After the autoxidation procedure was completed, the remaining amounts of ASA were determined using HPLC under the following conditions: column, LiChrosorb-NH2; mobile phase, acetonitrile-water-acetic acid (50:20:2, v/v); flow rate, 1.4 ml/min; UV detector (at 245 nm). The amounts of DHA yielded during autoxidation of ASA were measured according to the method described in literature using 2,4-dinitrophenylhydrazine reagent (2). Analysis of the oxidation products was carried out using GC and GC-MS. After the oxidation reaction was completed, the reaction mixture was evaporated to dryness also in the darkness, and 0.5 ml of TMSI-H (GL Sciences Inc., hexamethyldisilazane : trimethylchlorosilane : pyridine = 2:1:10) or 0.25 ml of N-methyl-N-trimethylsilyl-trifluoro-acetamide (MSTFA) was added to the residue to prepare TMS derivatives of the oxidation products. A Shimadzu model GC-9A gas chromatograph equipped with an FID detector was used for the analysis of TMS derivatives. An OV-1 capillary column (25 m x 0.25 mm i.d.) was used, and nitrogen was used as the carrier gas at the the flow rate of 0.9 ml/min. The column oven temperature was held at 60℃ for 5min then programmed to 210℃ with ascending temperature at a rate of 5 ℃/min. The structural information of these TMS derivatives was mainly obtained using a GC-MS system comprised of JEOL model JMS-DX300 mass spectrometer equipped with Hewlett Packard model 5790 gas chromatograph. The electron impact ionization method at 70 eV was used to ionize TMS derivatives, with the ion source temperature maintained at 180℃.

## Results and Discussion

Since ASA is water soluble, it has always been believed that the antioxidant effect of ASA is confined to the hydrophilic regions of food and biological systems, and that ASA molecules do not participate in the oxidation-reduction process occurring in hydrophobic regions of those systems.    Food and biological systems, however, usually have many constituents, and are essentially of multi-phase and multi-component nature.  Generally speaking, therefore, the solubility of ASA and oxygen depends on the environment (hydrophobic or hydrophilic), as shown in Figure 1.  Thus, highly water soluble ASA molecules might play some roles in the oxido-reduction reactions occurring in more hydrophobic regions than water.  From this point of view, further evaluation is needed for the chemical behavior of ASA molecules in solvents of lower dielectric constants and higher oxygen concentrations than water.

A preliminary evaluation was done for autoxidation rates of ASA in both aqueous and non-aqueous solvent systems.  As shown in Figure 2, ASA was quite unstable in MeOH solution where the concentration of oxygen was much higher than the aqueous solution, and the non-dissociated form of ASA in HCl solution was more stable than the dissociated one in water. Thus the autoxidation rates were strongly influenced by dissolved oxygen concentrations and ASA monoanion concentrations, while the non-dissociated form of ASA was quite stable against oxygen. Their rate constants are given in Table  I .

Separation and identification of initial autoxidation products of ASA in MeOH were carried out, and formation of DHA was positively confirmed as the main oxidation product of ASA.    Threonolactone was identified as its TMS derivative, 2,3-di-O-trimethylsilyl-L-threonolactone, by comparing its GC retention time with that of the authentic compound, and further confirmed by comparing the mass spectrum of the sample with the authentic mass spectral data reported in the literature (*3*).    The formation of oxalic acid was similarly confirmed by GC and GC-MS analyses. These reaction products were also detected in the autoxidation of ASA monoanion as an ASA-Na salt solution in MeOH, and the formation of the same oxidation products was confirmed in the autoxidation of ASA in an aqueous solution. It was also confirmed, however, that  these autoxidation reaction products of ASA were not formed from DHA.  Therefore, this appears to be a new autoxidation pathway of ASA that does not proceed via DHA, and this pathway might be involved in various oxidation processes observed in food and biological systems.

These experimental data obtained in the autoxidation of ASA in MeOH solution strongly suggested that, without the potent catalytic effect of heavy metal ions, ASA reacts with rather stable triplet oxygen to give very similar reaction products to those obtained in the reaction of ASA with singlet oxygen (*4, 5*), suggesting the formation of oxygen adduct of ASA monoanion as shown in Figure 3.

In order to obtain more detailed information regarding the autoxidation mechanism of ASA, semi-empirical molecular orbital calculations on the interaction of ASA monoanion and oxygen molecule were carried out using a general molecular orbital

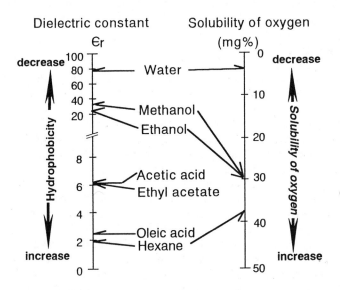

Figure 1 Solubility of oxygen and dielectric constant of various solvent

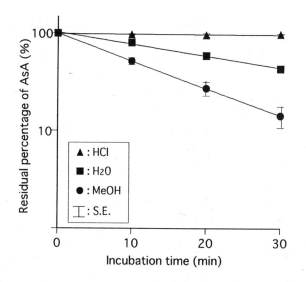

Figure 2 Autoxidation of AsA in various solution. Incubation temperature : 35℃

Table I The rate constants of AsA autoxidation

$(min^{-1})$

|  | 25°C | 30°C | 35°C |
|---|---|---|---|
| in water | 0.0083 | 0.0134 | 0.0283 |
| in methanol | 0.0251 | 0.0640 | 0.0753 |
| in HCl | 0.0002 | 0.0005 | 0.0009 |

AsA concentration : 0.03mg%

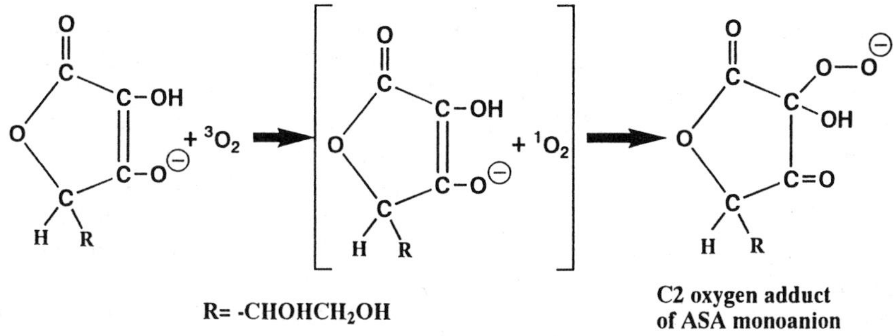

R= -CHOHCH₂OH

C2 oxygen adduct
of ASA monoanion

Figure 3 Reaction process of $^3O_2$ with ascorbate monoanion

package program, MOPAC (6). The highest occupied molecular orbital (HOMO) of ASA anion was found to be localized on C2 atom as has already been pointed out by previous studies (7,8). When the MO calculation was done assuming the distance between C2 atom of ASA and oxygen molecule as fixed at 10 Å, the results show that HOMO would be localized on C2 atom of ASA anion, while the lowest unoccupied molecular orbital (LUMO) would be localized on the oxygen molecule. If the oxygen molecule is present as singlet, its approach to C2 atom of ASA monoanion would cause a significant increase in the negative net charge on the oxygen molecule, and decrease in the heat of formation of the super-molecule of ASA anion with $O_2$, suggesting a strong interaction between these two molecules. When the distance between C2 atom of ASA monoanion and oxygen molecule became less than 1.5 Å, C2 oxygen adduct of ASA monoanion (ascorbate-2-peroxy-anion type intermediate) would be formed, and the optimized structure of this intermediate compound would show the bond length of C2-$O_2$ to be 1.42 Å. The optimized structure as well as HOMO and LUMO of this characteristic oxygen adduct are presented in Figure 4. Since the spin multiplicities of ASA anion and that of $^3O_2$ were different at their ground states, their direct reaction is so-called spin forbidden. However, as the distance between C2 atom of ASA anion and $^3O_2$ decreases, the difference in heat of formation between the super-molecules with ASA anion and $^1O_2$ and that with ASA anion and $^3O_2$ tends to decrease,

i.e., the apparent required energy for the excitation of $^3O_2$ to $^1O_2$ tends to decrease. In the vicinity of C2 atom of ASA anion, therefore, the needed energy to convert dioxygen to singlet oxygen, that is, relief of the spin restriction would be expected to be much reduced from the value of 23.4 kcal/mole which is usually taken as the activation energy required for this conversion process (9), suggesting the increased possibility for $O_2$ molecule to change its spin state from triplet to singlet. Thus, it is hypothesized that a triplet dioxygen molecule would come to behave like a singlet oxygen when the dioxygen molecule approached C2 atom of ASA anion very closely (Figure 3). These calculation results appear to support our experimental results described above.

It was reported that superoxide anion was formed in the oxidation of ASA in an aqueous solution, and the superoxide was further involved in the oxidation of ASA (10). Our previous study agreed to this finding and it was further confirmed in our experiment during the autoxidation of ASA in MeOH by using nitro blue tetrazolium reagent (11) (data not shown). From the results of MO calculations, superoxide anion might be produced by direct release from the peroxy anion type intermediate, or more likely from the dianion form of the peroxy anion type intermediate. Further investigations, however, are by all means needed to establish the mechanism of formation of superoxide anion.

Both threonolactone and oxalic acid were C2-C3 fission products, and their formation via the intermediate having a cyclic peroxide structure like dioxetane was considered to be the most probable mechanism as deduced from the reaction of singlet oxygen with ASA (4,5). As shown in Figure 4, HOMO and LUMO of the oxygen adduct were localized on one of the oxygen atoms of the peroxy anion moiety and C3 carbonyl group, respectively, and dioxetane structure would be easily formed by the reaction of the oxygen atom with C3-carbonyl carbon atom.

Thus, the initial autoxidation process of ASA monoanion with oxygen molecule is

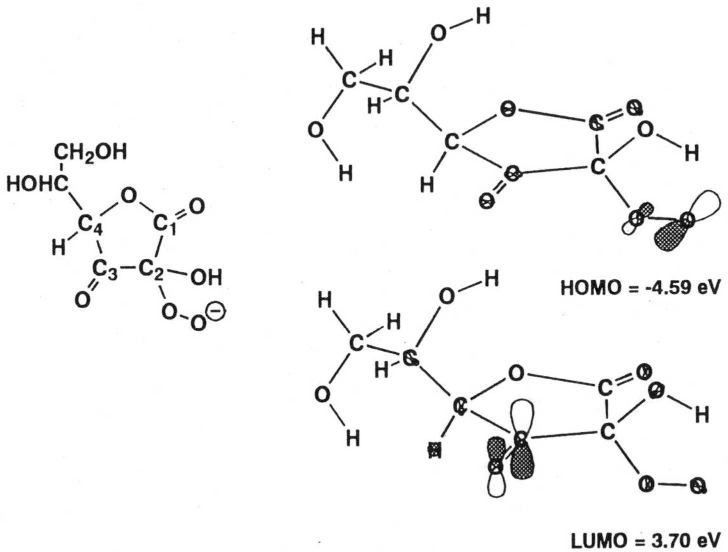

HOMO = -4.59 eV

LUMO = 3.70 eV

**Figure 4  HOMO and LUMO of ascorbate-2-peroxy-anion**

considered to proceed via the formation of oxygen adduct of ASA monoanion. Two different degradation pathways of this adduct appear to be present. The one is the release of superoxide anion and the concomitant formation of monodehydro-L-ascorbate anion radical which is known to easily undergo disproportionation reaction to form ASA and DHA. The other is the cleavage reaction of the bond between C2 and C3 of the adduct to yield considerably stable non-radical products, such as threonolactone and oxalic acid. We consider that the former pathway, the release of superoxide anion which would undergo protonation to the hydroperoxy radical followed by disproportionation to hydrogen peroxide, somehow related to the pro-oxidant effect of ASA. The latter pathway, cleavage of C2 and C3 of the adduct, is also considered related to the antioxidant effect of ASA. Dehydro-L-ASA (DHA), the main oxidation product of ASA, is an unstable compound having a characteristic tri-carbonyl system, and its optimized structure, charge distribution, and HOMO and LUMO are shown in Figure 5. In aqueous solution, DHA is considered to be very easily hydrated at C2 carbonyl group to give mono-hydrated DHA which appears to undergo a cyclization reaction forming a hemiketal ring (*12*). Alternatively, in aqueous solution, DHA is easily hydrolyzed to give DKG which is further degraded via the formation of 3,4-ene-diol form (*13*) and 2,3-ene-diol form (*14*) of its $\delta$-lactones to produce various types of carbonyl compounds (*15*) resulting in discoloration. Based on these findings, we consider DKG as an important intermediate compound in the browning reaction of ASA without involvement of amino compounds under the oxidative conditions. Without hydration and hemiketal ring formation, DHA is a typical dicarbonyl compound which easily reacts with amino compounds, such as amino acids. The reaction of DHA with an $\alpha$-amino acid is the so-called "Strecker Degradation", and by

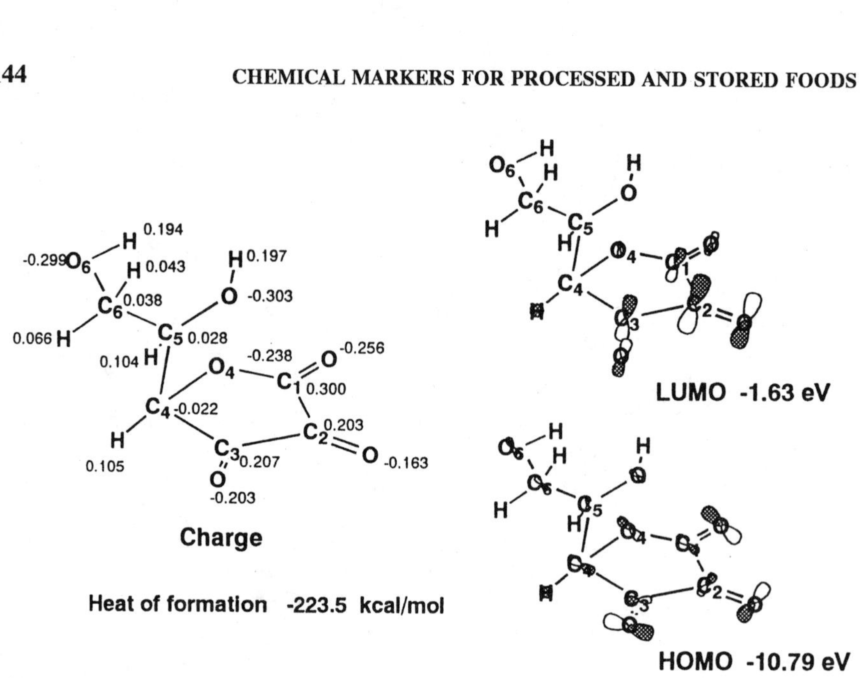

**Charge**

**Heat of formation** -223.5 kcal/mol

**LUMO** -1.63 eV

**HOMO** -10.79 eV

Figure 5  HOMO, LUMO, charge distribution of dehydro-L-ascorbic acid

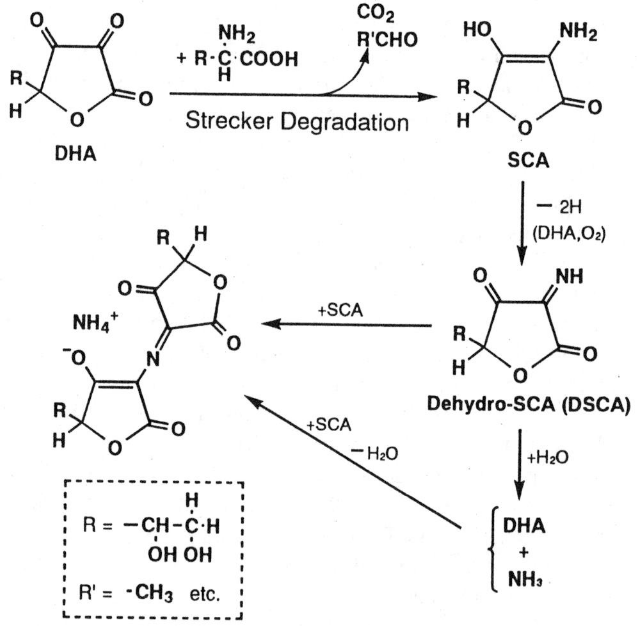

Figure 6  Formation of L-scorbamic acid and red pigment, 2,2'-nitrilodi-2(,2')-deoxy-L-ascorbic acid mono-ammonium salt (NDA)[15)]

this reaction process, DHA appears to yield SCA, an enaminol compound (*16*), which is a very important reaction intermediate in the formation of a red pigment with a characteristic chromophore as shown in Figure 6, and the red pigment itself is unstable, thus easily causes discoloration resulting in brown color.   SCA, therefore, is considered to be one of the important intermediates involved in the browning reaction of citrus juices of high ASA contents.

## Summary

The initial autoxidation process of ASA monoanion with oxygen molecule in the absence of metals is considered to proceed via the formation of oxygen adduct of ASA monoanion. After this initial autoxidation process, degradations of DHA and DKG take place and reactions of these carbonyl compounds with amino compounds also occur to give intensive discoloration.   Thus, autoxidation of ASA is considered to play an important role as a trigger reaction of discoloration, resulting in deterioration of various foods.

## Literature Cited

1.   von Euler, H.; Hasselquist, H. *Chem. Ber.* **1955**, *88*, 991-995.
2.   Roe, J. H.; Mills, M. B.; Oesterling, M. J.; Damron, C. M. *J. Biol. Chem.*, **1948**, *174*, 201-208.
3.   MacLafferty, F. W.; Stauffer, D. B., "The Wiley/NBS Registry of Mass Spectral Data", vol. 1-7, Wiley-Interscience (1974)
4.   Kwon, B.-M.; Foote, C. S. *J. Am. Chem. Soc.*, **1988**, *110*, 6582-6583.
5.   Kwon, B. -M; Foote, C. S.; Khan, S. I. *J. Am. Chem. Soc.*, **1989**, *111*, 1854-1860.
6.   Stewart, J. J. P., MOPAC version 6.0, Frank J. Seiler, Research Laboratory, U. S. Air Force Academy, Colorado, U. S. A.
7.   Abe, Y.; Okada, S.; Horii, H.; Taniguchi, S. *J. Chem. Soc. Perkin Trans. II.* **1987**, 715-720.
8.   Abe, Y.; Okada, S.; Nakano, R.; Horii, T.; Inoue, H.; Taniguchi, S.; Yamabe, S. *J. Chem. Soc. Perkin Trans. II.* **1992**, 2221-2232.
9.   Miller, D. M.; Buettner, G. R.; Aust, S. D. *Free Radical Biol. Med.*, **1990**, *8* 95-108.
10.   Scarpa, M.; Stevanato, R.; Viglino, P.; Rigo, A. *J. Biol. Chem.*, **1983**, *258*, 6695-6697.
11.   Beauchamp, C.; Fridovich, I.; *Anal. Biochem.*, **1971**, *44*, 276-287.
12.   Torbert, E. M.; Ward, J. B. "Dehydroascorbic Acid", in '*Ascorbic Acid; Chemistry, Metabolism, and Uses*' Seib, P. A.; Tolbert, B. M. (eds.), American Chemical Society, Washington, D.C., 1982; pp101-123.
13.   Otsuka, M.; Kurata, T.; Arakawa, N. *Agric. Biolo. Chem.*, **1986**, *50*, 531-533.
14.   Tanaka, H.; Kimoto, E. *Bull. Chem. Soc. Japan.* **1990**, *63*, 2569-2572.
15.   Kurata, T.; Fujimaki, M. *Agric. Biol. Chem.* **1976**, *40*, 1287-1291.
16.   Kurata, T.; Fujimaki, M., *J. Agric. Food Chem.*, **1973**, *21*, 676-680.

# Chapter 14

# Localized Antioxidant Degradation in Relation to Promotion of Lipid Oxidation

Marilyn C. Erickson[1], Robert L. Shewfelt[1], Brenda A. del Rosario[1], Guo-Dong Wang[1], and Albert C. Purvis[2]

[1]Center for Food Safety and Quality Enhancement, University of Georgia, Griffin, GA 30223
[2]Department of Horticulture, Coastal Plain Station, University of Georgia, Tifton, GA 31793

Quality deterioration in biological tissues emanating from lipid oxidation is restricted to localized areas and as such, we advocate that chemical measurements should be applied to those areas to accurately predict the loss of product quality. Localized oxidative degradation in plant and animal tissues is influenced by differences in metabolic state and cell and tissue structure. Degradation of antioxidants occurs in early stages of oxidation; consequently, measurement of antioxidant concentrations in specific tissue fractions over time has been used as a marker for the onset of a product's physical symptoms. Advantages and limitations inherent to the use of such a measurement have been investigated in relation to chilling injury of bell peppers. Other compositional factors that impact on the usefulness of localized antioxidant degradation as a marker of stored tissue quality will be introduced.

Lipid oxidation has been proposed as the vehicle for manifestation of injury and loss of quality in food by physiological stresses such as chilling injury, senescence and dehydration. The objective of this chapter is to review how localized degradation of antioxidants has been used as a marker for early detection of lipid oxidation as well as in defining the mechanism of oxidative challenge.

Chilling injury may be defined as the physiological damage that occurs in many tropical and subtropical species upon prolonged exposure to low but non-freezing temperatures (0-15°C). Several mechanisms have been postulated to explain physiological stresses such as chilling injury. These include altered calcium metabolism (1), changes in the cytoskeleton (2) and modifications of cellular membranes (3). Modifications of cellular membranes may in turn be accomplished by several routes: membrane lipid phase changes (4), modification of specific lipids within the membrane (5, 6), direct effects on membrane bound enzymes (7) and free radical degradation of membrane lipid (8-10). In our laboratory, we consider that free radical degradation is the most promising mechanism to explain chilling injury and have therefore directed our efforts towards testing that hypothesis. To reliably test that hypothesis, however, two precautions must be taken. First, the measurement(s) used must be sensitive to detect peroxidation in early stages so that it may be possible to differentiate between a cause and a symptom. Second, the measurement must be directed at a specific subcellular membrane rather than the entire tissue to avoid dilution of the oxidative response.

A localized oxidative response is a valid possibility when one considers how variable the composition of subcellular membranes can be. The ratio of protein to lipid in bell pepper mitochondria, cauliflower microsomes and bell pepper microsomes has been found to be 2.5, 3.9 and 3.3, respectively. The more dramatic differences, though, were found in the composition of tocopherols in these membranes. In pepper mitochondria, levels of tocopherol were one fourth the levels of lipid, in pepper microsomes they were one fifth and in cauliflower microsomes, one hundred times less tocopherol than lipid was present.

Differences in membrane composition can be reflected in the membrane's susceptibility to oxidation. For instance, when bell pepper microsomes and cauliflower microsomes were exposed to an iron-ascorbate oxidative challenge, a rapid increase in thiobarbituric acid reactive substances (TBARS) occurred within the first 30 min of incubation for both samples and little subsequent increase during the remaining 60 min (Figure 1A). At any time point, though, the level achieved in pepper microsomes was lower than it was in the cauliflower microsomes. Peroxidation in bell pepper mitochondria, in turn, was substantially lower than either of the microsomal samples. The relative susceptibility of these membranes to peroxidation, as measured by TBARS, appears to be in inverse proportion to the relative amount of tocopherol present in the membrane. Monitoring endogenous alpha-tocopherol levels initially, therefore, will often help to gauge the relative susceptibility of one's membranes.

Monitoring tocopherol throughout the oxidative challenge, though, can also prove useful (Figure 1B). Compared to their initial levels, tocopherol decreased gradually in both pepper mitochondria and cauliflower microsomes during their first 60 min of incubation, then decreased rapidly in the next 30 min. For pepper microsomes, tocopherol dropped to 85% of their initial levels by 60 min but did not drop further with extended incubation. Using both tocopherol and TBARS data, one interpretation of these data may be that during the first 30 min, particularly for the microsomal fractions, the primary event is breakdown of pre-existing hydroperoxides by reduced iron. This type of activity would explain the limited loss of tocopherol in these membrane fractions during this period because the alkoxyl radicals formed would have been less effectively trapped by tocopherol than would be the peroxyl radicals. This example also points out where tocopherol measurements used alone are ineffective as a monitor of oxidative stress but used with other measures provides useful mechanistic information.

Measurement of tocopherol will not always serve as an effective monitor of oxidative stress. The availability of other antioxidants, presence of free fatty acids and source of initiation have all been shown to affect tocopherol's effectiveness and levels. Examples of other antioxidants affecting tocopherol levels include the

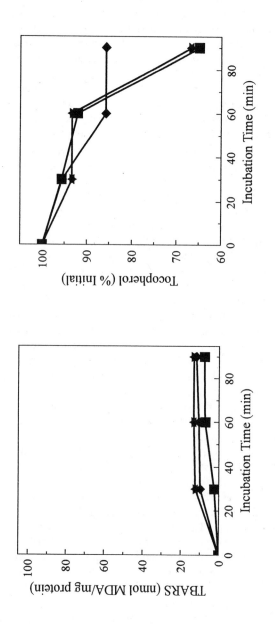

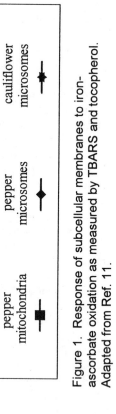

Figure 1. Response of subcellular membranes to iron-ascorbate oxidation as measured by TBARS and tocopherol. Adapted from Ref. 11.

protection by superoxide dismutase in mitochondrial/microsomal systems (*11*), the regeneration of tocopherol by ascorbic acid (*12*, *13*) and the potentiation of tocopherol's protective effect by glutathione (*14*) which is illustrated in Figure 2. In this investigation, initial tocopherol levels were maintained longer in the presence than in the absence of GSH and this extension corresponded to the extension in the lag phase for chemiluminescence, the measure of oxidation. By treating the microsomes with N-ethylmaleimide, which blocks the protein thiol groups, the ability of GSH to delay peroxidation was attenuated eliminating the possibility that GSH acted primarily through regeneration of tocopherol. The investigators also tested the hypothesis that GSH regenerated the protein thiol groups. So when GSH was added to the incubation system and the protein thiol groups were examined, the loss of protein thiols was slowed compared to the immediate loss of protein thiols in the absence of GSH. The rate of protein thiol loss in the absence of GSH also did not vary significantly during incubation in contrast to the successive slow and fast periods seen with lipid peroxidation and vitamin E loss. The investigators attributed the difference between the temporal patterns of thiol loss versus tocopherol loss to the likelihood that thiol groups of proteins, unlike $\alpha$-tocopherol, have a range of oxidation potentials, availabilities and reactivities, and so the most labile of these may react more readily than $\alpha$-tocopherol, while thiols lost at later times may represent less reactive residues. In any event, the results support the concept that protection of biological membranes can be multifaceted and integrative in that protein thiols and their maintenance will also afford an antioxidant capacity to membranes in addition to that provided by tocopherol.

Presence of free fatty acids is yet another factor which has been proposed to increase tocopherol's effectiveness (*15*). In support of this suggestion, the antioxidant effectiveness of $\alpha$-tocopherol in SDS/linoleic acid micelles and in multilamellar and unilamellar vesicles prepared from soy phosphatidylcholine has been evaluated using oxygen consumption, malonaldehyde formation and ESR spin trapping as measurements. While lipid peroxyl radicals in the micelles reacted with $\alpha$-tocopherol several hundred times faster than with the peroxyl radicals in the unilamellar vesicles, the reactivity ratio was still too small to allow physiological levels of $\alpha$-tocopherol to protect biological membranes against oxidative damage. Insertion of free fatty acids, however, markedly improved the antioxidant effectiveness of tocopherol.

The third variable known to modify an antioxidant's effectiveness and hence its ability to be used as a chemical marker of oxidative stress is the source of initiation. One of the classic examples to illustrate this point was presented by Niki (*16*) in which free radicals initiated in the water phase could be scavenged by ascorbic acid. Initiation in the lipid phase, however, minimized the effectiveness of ascorbic acid since this antioxidant only had access to radicals generated at the surface.

The source of initiation has also been shown by Palozza et al. (*17*) to modify tocopherol's effectiveness as an antioxidant. Three types of initiation were studied on rat liver microsomes by these investigators and they included: enzymic peroxidation induced by NADPH, ADP and iron; the water-soluble azo compound

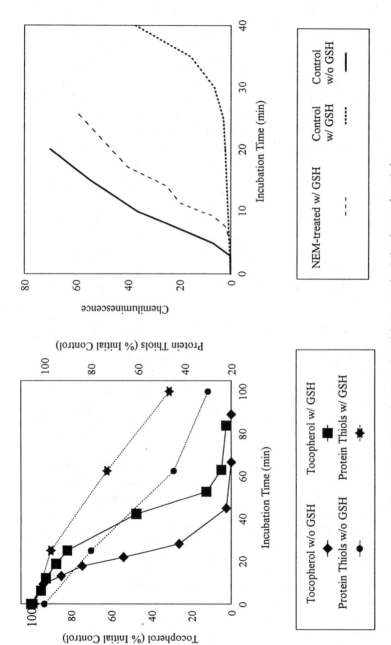

Figure 2. Protection by glutathione against the loss of protein thiols and tocopherol. Adapted from Ref. 14.

AAPH (2,2'-azobis(2-amidinopropane)); and the lipid soluble azo compound AMVN (2,2'-azobis(2,4-dimethylvaleronitrile)). In this study, they were interested in the degree of tocopherol loss prior to the production of TBARS. As the prooxidant varied, different temporal relationships existed between the loss of tocopherol and the observance of lipid peroxidation. A substantial depletion of about 70% of endogenous tocopherol preceded the propagation phase when induced by either of the azo compounds while only 20% of the antioxidant had disappeared before peroxidation began when induced by the enzymic system.

In our laboratory, we have also demonstrated that the source of initiation can affect plant membranes differentially (*18*). For cauliflower microsomes, xanthine oxidase leads to significantly greater degradation of tocopherol than the other two sources of initiation, iron-ascorbate or cumene hydroperoxide, whereas in bell pepper microsomes, tocopherol was equally affected by cumene hydroperoxide and xanthine oxidase (*11*).

In addition to having oxidative destruction of polyunsaturated fatty acids, free radical attack on biological membranes can lead to protein damage, through fragmentation, cross-linking and amino acid modification. In cases where there is an intimate association between the proteins and lipids, rather than a direct attack by some active oxygen species, many investigators envision that free radicals derived from lipid peroxidation are actually the culprit in damage to the proteins. Under such conditions, one may hypothesize that inhibition of lipid peroxidation would lead to inhibition of protein degradation. Confirmation of such a response was found by Dean and Cheeseman (*19*) for the fragmentation of the membrane protein monoamine oxidase. In non-supplemented particles, the percent protein degradation paralleled the increase in TBARS (Figure 3). When the rats were injected with $\alpha$-tocopherol prior to isolation of the membranes, the levels of this antioxidant in the membranes were increased ten fold. Under a similar oxidative challenge as applied to the control sample, both protein degradation and TBARS were inhibited in the tocopherol loaded sample.

Our initial investigations, in contrast, have not found a similar protection by tocopherol against proteolysis in plant mitochondrial membranes. For pepper mitochondria, the degree of proteolysis is quite large over a 90 min incubation period with iron-ascorbate (Figure 4) despite the higher initial levels of tocopherol in mitochondria than the microsomal fraction. Also, only between 60 and 90 min of incubation were significant losses of tocopherol noted in the mitochondrial fraction (Figure 1B) and this time period was long after significant proteolysis had occurred.

To identify if the inability of tocopherol to protect mitochondrial proteins is specific to the source of oxidative stress, xanthine oxidase, was also applied to the plant membranes. While proteolysis was not as severe for the mitochondrial membranes as it had been when iron-ascorbate was the initiating agent, the proteolysis in mitochondrial membranes still exceeded the proteolysis in either of the microsomal membrane fractions (Figure 5a). In addition, tocopherol loss does not precede proteolysis for the mitochondrial fraction and this was also the case for the microsomal fractions (Figure 5b). Based on these data and the preceding data for iron-ascorbate, it would appear that tocopherol is limited in its ability to

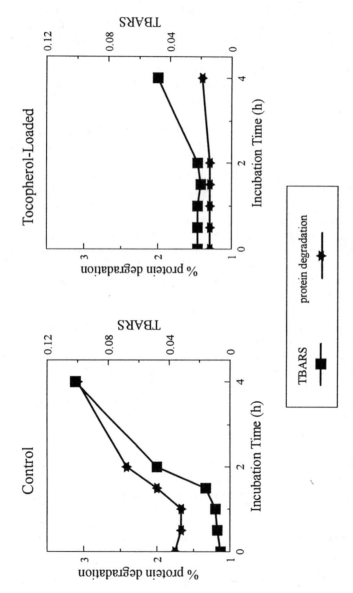

Figure 3. Protection of monoamine oxidase degradation by tocopherol. Adapted from Ref. 19.

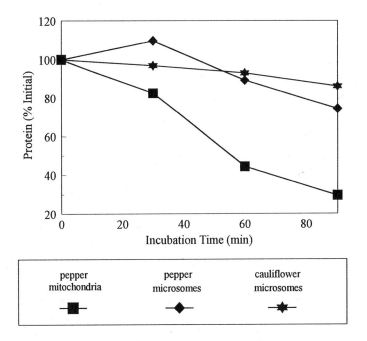

Figure 4. Proteolysis of plant subcellular membranes in response to iron-ascorbate oxidation. Adapted from Ref. 11.

protect plant membrane proteins. Based on some data that will be presented next, though, this statement cannot be applied to all plant membranes.

One of the problems with working with a microsomal fraction is that it represents a heterogeneous mixture of small vesicular membranes that originate to varying degrees from plasma membrane, endoplasmic reticulum, mitochondria and thylakoids. With such a mixture, there is still the possibility that if one particular membrane is affected by the oxidative stress, it would not be seen due to dilution by the nonaffected membranes. To avoid this possibility, our laboratory has chosen to work with isolated plasma membrane fractions due to the postulated role of this membrane in chilling injury through breakdown of the membrane bound transport enzyme $H^+$-ATPase (*20*).

In working with plasma membranes and its marker enzyme, $H^+$-ATPase, one must keep in mind that during isolation and reconstitution of the disrupted plasma membranes, varying degrees of inside-out vesicles form that depend on the composition of the reconstitution medium. For the medium used in our laboratory (*20*), 30% of the membrane vesicles will be inside-out making it now possible to measure the membrane bound enzyme $H^+$-ATPase because its active site is now oriented to the outside (Figure 6). For the remaining 70% of membranes which are right-side out, $H^+$-ATPase activity is not measurable if left intact since the ATP cannot access the active site of the enzyme.

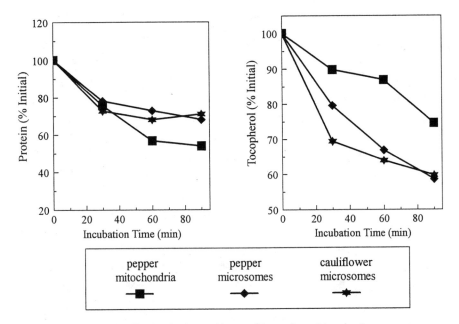

Figure 5. Proteolysis and loss of tocopherol in plant subcellular membranes in response to xanthine oxidase. Adapted from Ref. 11.

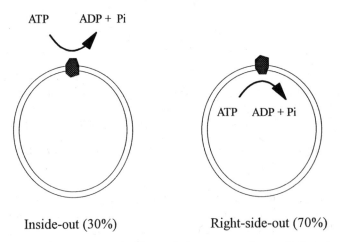

Figure 6. Vesicle sidedness of plasma membrane isolated from bell pepper fruit tissue.

Bearing this fact in mind, Figure 7 shows the effect of lipid peroxidation on intact plasma membrane vesicles. Two types of initiation systems were used: lipoxygenase-phospholipase which is capable of affecting the membrane $H^+$-ATPase only through the lipid; and iron-ascorbate capable of affecting the $H^+$-ATPase through oxidation of lipids but also by generating reactive oxygen species which could directly attack the $H^+$-ATPase enzyme. After 60 min of incubation, comparable levels of oxidative products were generated in both initiation systems but they both exhibited a lag period in the first 30 min of incubation. In contrast, $H^+$-ATPase activity was lost fairly uniformly throughout the incubation and to similar degrees with both initiation systems.

For Triton-solubilized plasma membranes, where both right-side out and inside-out $H^+$-ATPase can be measured, there was also no difference in loss of $H^+$-ATPase activity between the two types of initiation systems (Figure 8). These data together with the previous data from intact vesicles, suggested that $H^+$-ATPase activity was lost primarily through the action of lipid radicals.

If true, membrane antioxidants should be expected to protect $H^+$-ATPase. Some data that would lend support to this statement are presented in Figure 9. In the absence of any oxidative challenge, the $H^+$-ATPase activity increased in solubilized and intact vesicles. Oxidized sulfhydryl groups on the $H^+$-ATPase which arise during isolation may be regenerated by other reducing substances in the membrane but do so only after they have had time to migrate to the $H^+$-ATPase. The greater increase in $H^+$-ATPase activity seen in solubilized vesicles than in intact vesicles is reasonable to expect since solubilized vesicles would exhibit greater mobility.

Some indirect evidence to support that it is tocopherol which protects $H^+$-ATPase comes from some observations made during isolation of the plasma membrane. In the absence of dithiothreitol in any of the suspension buffers, smeared protein bands were observed on electrophoresis gels from plasma membranes as opposed to a clear $H^+$-ATPase protein band being observed from the microsomal fraction. Oxidative degradation of the plasma membrane proteins appeared to be occurring and this degradation in the microsomal fraction is not seen because of the significant loss of tocopherol, 82%, which occurred within the first 6 hours of storage. A similar lack of protection for $H^+$-ATPase by tocopherol in the plasma membrane may not occur due to insufficient levels of tocopherol being present. What is interesting is when dithiothreitol is added to the suspension buffer, oxidative degradation of the proteins does not occur in either the plasma membrane fraction or the microsomal membrane fraction. In addition, tocopherol levels are stabilized for up to 6 hours. Thus, the dithiothreitol may be reducing the oxidized sulfhydryl groups on the $H^+$-ATPase molecule and sparing the tocopherol which otherwise would be protecting the $H^+$-ATPase. More in depth evidence is needed, however, before this type of interaction can be conclusively stated. If there is such an interaction, plasma membrane tocopherol could prove useful as a marker for the onset of chilling injury.

In conclusion, monitoring antioxidant degradation in localized areas of tissue can prove useful as a marker for the onset of a product's quality loss. The multifaceted protection available to biological membranes, though, would suggest that measurements of more than one antioxidant are warranted during storage.

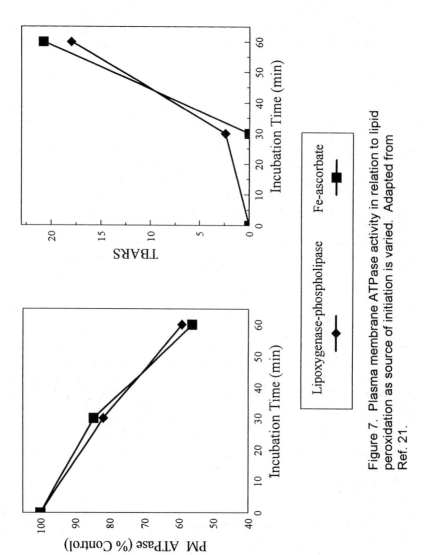

Figure 7. Plasma membrane ATPase activity in relation to lipid peroxidation as source of initiation is varied. Adapted from Ref. 21.

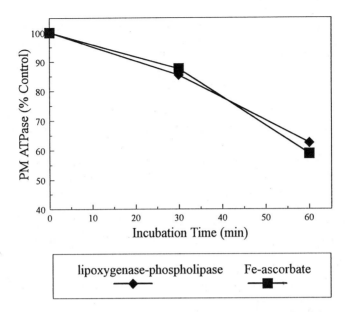

Figure 8. Solubilized plasma membrane ATPase activity in relation to source of initiation for lipid peroxidation. Adapted from Ref. 21.

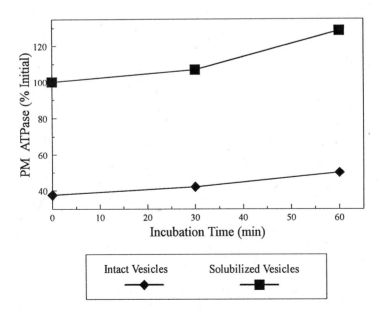

Figure 9. Plasma membrane ATPase activity in solubilized and intact vesicles following reconstitution. Adapted from Ref. 21.

While the examples presented in this chapter were from model system incubations, similar valuable information will be obtained by monitoring antioxidant degradation in specific membrane fractions from stored tissue.

**Literature Cited**

1    Minorsky, P.V. *Plant Cell Physiol.* **1985**, *24*, 81-86.
2.   Raison, J.K.; Orr, G.R. In *Chilling Injury of Horticultural Crops*; Wang, C.Y., Ed.; CRC Press: Boca Raton, FL, 1990; pp. 145-164.
3.   Steponkus, P.L *Ann. Rev. Plant Physiol.* **1984**, *35*, 543-584.
4.   Lyons, J.M. *Ann. Rev. Plant Physiol.* **1973**, *24*, 445-466.
5.   McKersie, B.D.; Hoekstra, F.A.; Krieg, L.C *Biochim. Biophys. Acta* **1990**, *1030*, 119-126.
6.   Whitaker, B.D. *Phytochem.* **1993**, *32*, 265-271.
7.   Thompson, J.E.; Legge, R.L.; Barber, R.F *New Phytol.* **1987**, *105*, 317-344.
8.   Winston, G.W. In *Stress Response in Plants: Adaptation and Acclimation Mechanisms;* Alscher, R.G. ; Cumming, J.R., Eds.; Wiley-Liss, Inc.: New York, NY, 1990; pp. 57-86.
9.   Shewfelt, R.L.; Erickson, M.C. *Trends Food Sci. Technol.* **1991**, *2*, 152-154.
10.  Purvis, A.C.; Shewfelt, R.L. *Physiol. Plant.* **1993**, *88*, 712-718.
11.  del Rosario, B.A. *Mitochondrial respiration and lipid peroxidation: Mechanistic studies in the development of chilling injury in bell pepper (Capsicum annum) fruit.* M.Sc. Thesis. University of Georgia, Athens, **1994**.
12.  Doba, T.; Burton, G.W.; Ingold, K.U. *Biochim. Biophys. Acta* **1985**, *835*, 298-303.
13.  Burton, G.W., Hughes, L., Foster, D.O., Pietrzaki, E., Goss-Sampson, M.A. and Muller, P.R. In *Free Radicals: From Basic Science to Medicine*; Poli, G.; Albano, E.; Dianzani, M.U., Eds.; Switzerland, Birkhauser Verlag Basel, 1993; pp. 388-402.
14.  Murphy, M.E.; Scholich, H.; Sies, H *Eur. J. Biochem.* **1992**, *210*, 139-146.
15.  Church, D.F.; Dugas, T.D.; Blazier, J.D. *Inform* **1995**, *6*, 480.
16.  Niki, E. *Chem. Phys. Lipids* **1987**, *44*, 227-253.
17.  Palozza, P.; Moualla, S.; Krinsky, N.I. *Free Rad. Biol. Med.* **1992**, *13*, 127-136.
18.  Cowart, D.M.; Erickson, M.C.; Shewfelt, R.L. *J. Plant Physiol.* **1995**, *146*, 639-644.
19.  Dean, R.T.; Cheeseman, K.H *Biochem. Biophys. Res. Commun.* **1987**, *148*, 1277-1282.
20.  Wang, G-D.; Morre, D.J.; Shewfelt, R.L *Postharv. Biol. Technol.* **1995**, *6*, 81-90.
21.  Wang, G.-D. *Study of isolation of plasma membrane, characterization of plasma membrane $H^+$-ATPase and effect of in vitro peroxidation on plasma membrane from bell pepper fruit tissue (Capsicum annum L.) fruit tissue.* M.Sc. Thesis, University of Georgia, Athens, **1992**.

Chapter 15

# Application of Enzymatic Reactions for Evaluation of Quality Changes in Frozen Vegetables

K. H. Park[1], J. W. Kim[2], and M. J. Kim[1]

[1]Department of Food Science and Technology and Research, Center for New Bio-Materials in Agriculture, Seoul National University, 103 Seodun-Dong, Kwonsun-Ku, Suwon 441−744, Korea
[2]Department of Biology, University of Inchon, Inchon 402−749, Korea

Thermal inactivation of lipid–acyl–hydrolase (LAHase) which is capable of hydrolyzing phospholipid and galactolipid was investigated to optimize blanching process prior to frozen storage of vegetables. The results of in–situ analysis, measuring decomposition of lecithin, mono- and digalactosyl diglyceride indicated that LAHase could be adequately used as the indicator enzyme for determination of quality deterioration of frozen vegetables. A time–temperature indicator (TTI) using a phospholipid–phospholipase system was also developed to monitor quality change of frozen foods. The TTI containing phospholipid, phospholipase, a mixture of pH indicators, and antifreeze reactants was designed for reactions at sub–zero temperatures. The TTI was more reliable than the lipase system in monitoring quality changes of frozen food during storage. The TTI had an activation energy of 32.1 kcal/mol, suitable for many food reactions.

Blanching of vegetables is generally evaluated by the residual activity of peroxidase in them (1). The blanching time should be long enough to destroy the indicator enzyme, peroxidase. However, this 'presence-absence' assay does not provide satisfying indication of adequate blanching. There is no evidence that the enzyme is directly related to specific quality deterioration of frozen vegetables (2). Peroxidase is one of the most heat stable enzymes in vegetables and heating vegetables to complete absence of peroxidase activity is more than adequate to inactivate other enzymes responsible for the reactions causing quality loss. Williams et al. (3) proposed to take lipoxygenase as the indicator enzyme for a

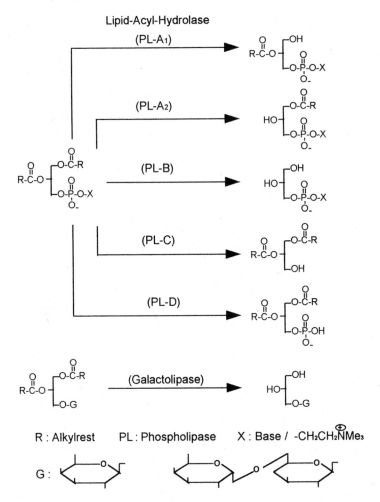

Figure 1.  Possible reactions of lipid–acyl hydrolases in plants.

routine assay to determine adequate blanching of green peas and green beans.   Barret and Theerakulkait (*1*) investigated the merits of using lipoxygenase as an appropriate blanching indicator instead of peroxidase.

Generally, triglyceride degrading lipases are not found often in green plants, while high activities of phospho- and galactolipases are present widely in them.   Their substrates, phospho- and galactolipid, are accordingly the major plant lipids as shown in Table I.   Therefore, the enzyme activities degrading the polar lipids seem to play an important role for quality determination of vegetables and fruits.

In glycolipids, glycerol is acylated at the C-1 and C-2 positions and linked glycosidically at C-3 to form a mono--, a di-, or a triglyceride.   In higher plants and algae, and particularly in the chloroplast of them, the sugar residue is galactose.   The associated fatty acids are highly unsaturated (*4*).

Table I.   Fat Soluble Components in Spinach Leaves (*3*)

| Total Lipid | 89 % | Total Pigments | 11 % |
|---|---|---|---|
| phospholipid | 17 % | chlorophyll | 9 % |
| galactolipid | 66 % | carotenoid | 2 % |
| neutral lipid | ~6 % | | |

The most common phospholipids are phosphatidyl esters in which phosphatidic acid is esterified with another hydroxy compound to form a diester of phosphoric acid.   Incorporation of choline and ethanolamine as the hydroxy compound would result in the formation of phosphatidyl-choline (lecithin) and phosphatidyl ethanolamine (cephaline), respectively. There are enzymes specifically hydrolyze one or more of the four ester bonds in phospholipids and/or galactolipid.   Each of the enzymes, phospholipase $A_1$ (PL-$A_1$), phospholipase $A_2$ (PL-$A_2$), phospholipase C (PL-C), and phospholipase D (PL-D), cleaves a phosphatidyl ester into two products as shown in Figure 1.

Lee and Mattick (*5*) observed degradation of phospholipid in unblanched pea during storage at -17.8 ℃.   Pendlington (*6*) also reported that the content of phospholipid decreased while that of choline and phosphate increased during the storage of unblanched pears.   Lea and Parr (*7*) described that fish taste caused by hydrolysis of phospholipid and galactolipid.   Oursel et al. (*8*) have observed the activity of phospholipase C in spinach leaves and demonstrated that addition of phospholipase C from *Bacillus cereus* to isolated chloroplasts increased the formation of

1,2-diacylglycerol and consequently, the formation of galactolipids by galactosylation. Hirayama and Oido (3) investigated changes of lipid and pigment contents in spinach leaves during storage. Lipid decreased more rapidly than pigment. They suggested that heat treatment at 100 ℃ stimulated phospholipase D activity but repressed galactolipase activity. Phospho- lipase D activity in flaked soybeans was assayed during storage and inactivation study of the enzyme was carried out using microwave heating (9). The enzyme was activated during the early stage of heating and then gradually destroyed till the temperature reached 115-120 ℃. When phospholipase D in soybean flakes was treated with live steam, the enzyme activity was rapidly destroyed at about 110 ℃.

The kinetics of heat stability of LAHases including phospholipase C, D, and galactolipase in plant tissues have not been determined systematically. In this study, we thermal stability of these enzymes for the application to thermal process of vegetables. The isolated enzymes could show greatly different stability from those present in the plant tissues since the cell components might play a role to destabilize the enzymes. We also developed a rapid in-situ test for the detection of the enzyme activity, which can be practically applied to foods.

## Materials and Methods

**Determination of LAHase Activity.** Freeze-dried spinach powder (0.05 g) was suspended in 2.5 ml of 0.1 M phosphate buffer solution (pH 7.0) and stirred with a magnetic bar at 30 ℃ for 20 or 60 min. The reaction was stopped by adding 2.5 ml methanol. The precipitate was collected by centrifugation and lipid was extracted from the precipitate with 3 ml of chloroform/methanol mixture (2 : 1). Aliquotes of the supernatant (2 ml) was dried under nitrogen stream and then 0.5 ml chloroform was added. The extract was applied onto a Kieselgel-G-plate (Merck) activated for 20 min at 110 ℃ and developed in a chromatography chamber with eluent (solvent : chloroform : methanol : water= 75 : 25 : 4). The contents of lecithin, monogalactolipid, and digalactolipid were determined by in-situ method.

**Determination of Phospholipase C and D Activities.** Acetone powder of phospholipase C and D was obtained by treating spinach leaves with a same volume of acetone at -18 ℃. For phospholipase C, the procedures described by Kuroshima et al. (10) were used. Enzyme solution (0.1 ml) was mixed for several seconds with 0.1 ml of toluene, 1.8 ml of distilled water, 0.5 ml of Tris buffer (pH 7.5), and 0.6 ml of p-nitrophenyl-phosphorylcholine in a vial. After an hour of incubation at 30 ℃, 8 ml of ethanol was added to the mixture and the reaction solution was filtered through a filter paper.   Then, 2 ml of 2 M Tris solution was added and the yellow color intensity was measured spectrophotometrically at 400 nm. The content of p-nitrophenol was calculated by extrapolating OD values to a calibration curve.   One unit of enzyme is defined as the amount of enzyme that releases 1 μmole of p-nitrophenol per min at 30 ℃.

Phospholipase D activity was assayed based on spectrophotometric measurement of choline (*11*).   A small portion of a chloroform solution containing lecithin (10 μmole) was evaporated to dried state in a cortex tube under nitrogen gas, and then the dried lipids were resuspended in 2 ml of 50 mM Tris buffer (pH 7.5) by ultrasonicating the mixture for 60 seconds in an ice water bath.   The solution was mixed with 0.2 ml of 125 mM CaCl2 and 0.1 ml of enzyme solution to make a total volume of 2.5 ml. The reaction mixture was incubated at 37 ℃ for 15 min.   At the end of incubation, the reaction tube was cooled down immediately in an ice bath and then 0.2 ml of 20 % (w/v) BSA solution was added, followed by the addition of 0.5 ml of 20 % (v/v) perchloric acid solution.   After mixing the solution vigorously with a vortex mixer, the reaction tube was centrifuged for 10 min.   The supernatant was transferred to another tube, cooled down to 4 ℃, and 0.2 ml of cold potassium triiodide solution was added to obtain choline periodide precipitate.   After centrifugation, the supernatant was completely pipetted off the firmly packed pellet.   The choline periodide formed was immediately dissolved in 5 ml of 1,2-dichloroethane.   The absorbance of the solution was measured at 365 nm.

**Determination of Peroxidase activity.**   Suspension of freeze-dried spinach powder was allowed to stand at room temperature for 20 min and filtered through cheese cloth.   The filtrate (peroxidase extract) was fractionated with 40-85 % ammonium sulfate and the precipitate was dissolved in 0.1 M phosphate buffer (pH 7.0).   The isolated enzyme solution was freeze-dried after dialysis against water.   The enzyme activity was measured as described previously (*12*).

**Thermal Inactivation of Enzymes.**   Freeze-dried spinach powder (0.05 g) was rapidly dissolved in 2.5 ml of 0.1 M phosphate buffer (pH 7.0) in a flask, which was equilibrated to a desirable temperature in a water bath, by stirring with a magnetic bar.   The temperatures of the samples reached the inactivation temperatures very rapidly due to   vigorous stirring.   The sample solution was incubated for a period of 0.5 – 60 min, cooled down to 30 ℃, and subjected to further reaction.   For inactivation of isolated enzyme, 0.2 ml of enzyme solution was rapidly introduced into 2.8 ml of the same buffer as described above.

## Results and Discussion

**in-situ Measurement of Residual LAHase Activity.**   The suspension of spinach was incubated for 20 or 60 min and then applied onto a TLC

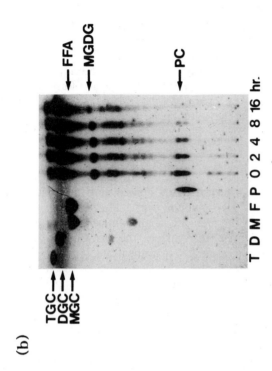

Figure 2 Thin layer chromatography of lipid acyl hydrolase activi in spinach leaves. In panel a, a suspension of dried spinach leav were incubated at 30 °C for 0, 20, or 60 min with or without heat treatment (60 °C) before applied onto a TLC plate. In panel b, the suspension was incubated at 30 °C for 0–18 hours. Triglyceride (TGC or T), diglyceride (DGC or D), monoglyceride (MGC or M), free fatty acids (FFA or F), and phosphatidylcholine (PC or P) were used as controls.

plate (Figure 2a). The concentration of substrates, lecithin, di- and monogalactolipid, decreased greatly in the spinach samples not treated with heat while that remained constant in the heat treated samples. This indicated that disappearance of the substrates was due to enzymatic hydrolysis in the extract. We observed that the decrease of substrates exhibited linearity as the reaction time increased within 60 min so that the enzyme activity could be estimated from the time course curve. The extract was applied to TLC and developed with a solvent containing diethylether/hexane/ammonia (50:50:0.25) upto 18 cm above the start line, chloroform/methanol/water/ammonia (70:30:3:2) 10 cm further, and ethyl-acetate/acetone/water/glacial acetic acid (40:40:2:1) to 16 cm (*13*). Figure 2b showed that the amount of free fatty acids increased as the reaction proceeded while those of mono-, digalactolipids, and phosphatidylcholine (lecithin) decreased.

**Inactivation of LAHase.** In Figure 3, decrease of phospho-, mono-, and digalactolipase activities was plotted semi-logarithmically at various heating temperatures. The residual activity curves in the range of 45-60°C indicated that thermal inactivation of phospho-/galactolipid degrading enzyme deviated from first order kinetics. All of them are biphasic kinetic curves. However, the enzyme was inactivated mainly during early stage of heating, and the D-values (decimal reduction time) taken from the initial steeper slope are shown in Table II.

Table II. Thermal Inactivation of LAHases in Spinach Leaves at pH 7.0

| | D-value (s) | | | | | |
|---|---|---|---|---|---|---|
| Enzymes | 45°C | 50°C | 55°C | 60°C | 70°C | $z$ (°C) |
| Lecithinase | 2400 | 1140 | 372 | 72 | | 9.8 |
| Digalactolipase | 1500 | 540 | 174 | 42 | | 10.0 |
| Monogalactolipase | 2640 | 1200 | 768 | 84 | | 9.0 |
| Isolated phospholipase | | | | | | |
| Phospholipase C | 4960 | 3237 | 2870 | 850 | 350 | 20.8 |
| Phospholipase D | | 1804 | 317 | 133 | 9.72 | 8.8 |

Source : reprinted from rf. 22

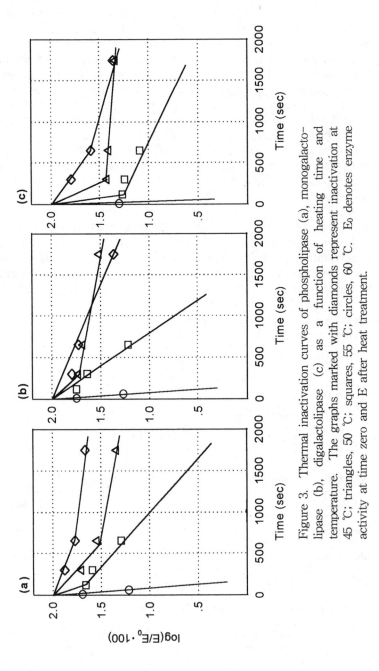

Figure 3. Thermal inactivation curves of phospholipase (a), monogalacto-lipase (b), digalactolipase (c) as a function of heating time and temperature. The graphs marked with diamonds represent inactivation at 45 °C; triangles, 50 °C; squares, 55 °C; circles, 60 °C. $E_0$ denotes enzyme activity at time zero and E after heat treatment.

Thermostability of lecithinase and monogalactolipase was slightly greater than that of digalactolipase. However, the dependence of D-value on temperature seems to be parallel, so that they provided very similar z-values (Figure 4 and Table II). Lipid acyl hydrolase from cowpea was purified to homogeneity and exhibited the hydrolytic activity on different substrates in the order of DGDG > MGDG > PC to produce free fatty acids (14). Thus the enzyme appeared to be a non-specific acylhydrolase, acting on galactolipids as well as on phospholipids. Phospholipases in spinach have been known to be mainly C or D types. Oursel et al. (15) have partially purified phospholipase C from spinach leaves. Acetone powder of phospholipase C and D was obtained by treating spinach leaves with acetone at -18 ℃. The activity of phospholipase C in spinach was detected. The acetone powder of spinach was incubated with egg lecithin at 30 ℃ and then the lipids were extracted with chloroform/methanol (3 : 1) from the reaction mixture. The chromatogram showed that diglyceride was produced (data not shown). In order to compare thermostability of phospholipases, the isolated and the intracellular forms, we isolated phospholipase C and D from spinach leaves and investigated their thermostability. Phopholipase C was more stable than lecithinase (Figure 5). If we assume that the lecithinase activity would consisted of phospholipase C, the enzyme was more stable when isolated from spinach leaves than when present in the intact leaves. This difference is probably due to the effect of cell components. The discrepancy was similar to the case of peroxidase in that isolated peroxidase was more stable in buffer solution than in the presence of cellular components (12). This discrepancy indicated that certain model experiments cannot be directly applied to foods.

**Thermal Inactivation of Peroxidase.** To study thermostability of spinach peroxidase without the influence of cellular components, isolated peroxidase was heat treated in 0.1 M phosphate buffer (pH 6.0). The thermal inactivation curves are presented in Figure 6 and they showed biphasic kinetic curves in the range of 60-70 ℃. The spinach suspension and the extract (Figure 6b and c) also exhibited biphasic curves similar to that of isolated peroxidase (Figure 6a). The thermodynamic data were summarized in Table III. The results indicated that isolated peroxidase was more thermostable than those in the suspension or in the extract in the range of 50-80 ℃. However, z-value was 13 ℃ for isolated peroxidase which was less than those (~18 ℃) for both of the enzymes in the suspension or the extract. The results showed that peroxidase isolated from spinach responded differently to heating than the enzymes in the spinach extract or suspension. In Table IV, heat stabilities of peroxi-

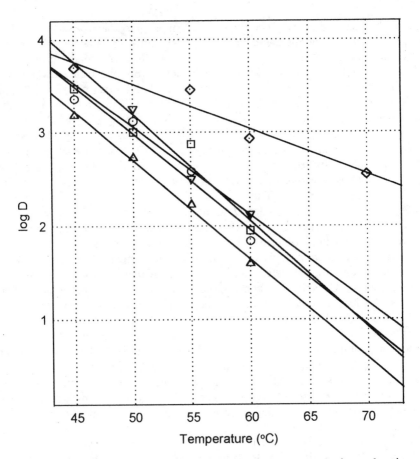

Figure 4.    D-values of lipid acyl hydrolase at various heating temperatures.    The graph marked with circles represents phospholipase; triangles, digalactolipase; squares, monogalactolipase; diamonds, phospholipase C; and reversed triangles, phospholipase D.

dases from various plants were summarized.    The $z$-value of spinach peroxidase was 13 °C which was less than that of horseradish.    Park et al. (16) investigated the influence of milieu factors on thermal inactivation of peroxidase.    The most effective were lecithin and monoglyceride in inactivating horseradish peroxidase.    In the presence of lecithin, peroxidase was inactivated at temperatures as low as 0 °C and pH 4.0.

Table III. Thermal Inactivation of Spinach Peroxidase

| Enzymes | pH | D-values (s) | | | | | z-values |
|---------|-----|------|------|------|------|------|----------|
| | | 50℃ | 60℃ | 70℃ | 80℃ | 90℃ | (℃) |
| Isolated | 6.0 | | 2560 | 744 | 79.8 | 14 | 13 |
| Extract | 4.0 | | 280 | 100 | 20 | | 18.3 |
| | 5.0 | | 702 | 200 | 40 | | 16.8 |
| | 5.5 | | 500 | 162 | 40 | | 18.3 |
| | 6.0 | | 600 | 120 | 40 | 5.8 | 18 |
| | 7.0 | | 282 | 60 | 19.8 | | 18 |
| | 8.0 | 420 | 198 | 30 | | | 18 |
| Suspension | 6.0 | | 840 | 282 | 60 | | 17.5 |

Source : reprinted from rf. 12

Table IV. Heat Stability of Peroxidases in Plants (*17, 18, 19*)

| Peroxidases in | | D-value (s) | | | | z-value |
|----------------|---|------|-------|------|------|---------|
| | | 60℃ | 70℃ | 80℃ | 90℃ | (℃) |
| Papaya | | 2505 | 485 | 50 | | 12 |
| Horseradish I | | | 2255 | 594 | 333 | 24 |
| | II | | 14500 | 1850 | 400 | 12.5 |
| | III | | 10700 | 2050 | 875 | 18 |
| | IV | 567 | 264 | 78 | | 23.7 |
| Spinach | | 2560 | 744 | 80 | 14 | 13 |
| Apple | | | 660 | | | |

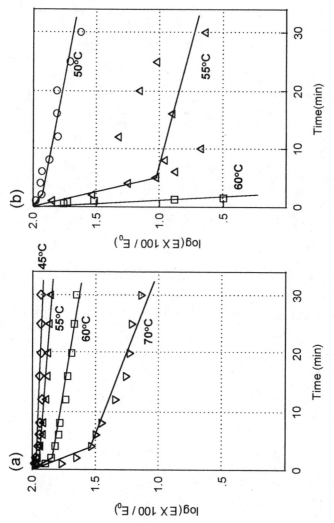

Figure 5.  Thermal stability of phospholipase C (a) and D (b).

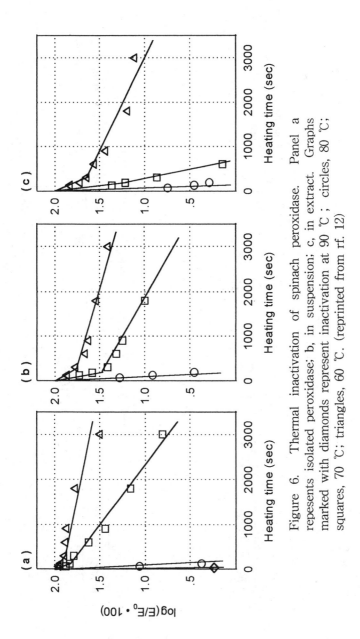

Figure 6. Thermal inactivation of spinach peroxidase. Panel a repesents isolated peroxidase; b, in suspension; c, in extract. Graphs marked with diamonds represent inactivation at 90 °C ; circles, 80 °C; squares, 70 °C; triangles, 60 °C. (reprinted from rf. 12)

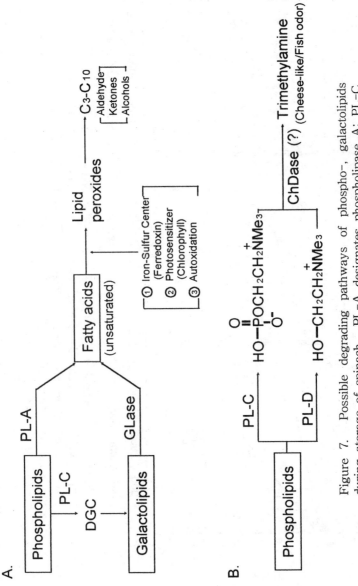

Figure 7. Possible degrading pathways of phospho-, galactolipids during storage of spinach. PL-A designates phospholipase A; PL-C, phospholipase C; DGC, diglyceride; GLase, galactolipase; PL-D, phospholipase D; ChDase, choline deaminase.

Galactolipid and/or Phospholipid Degrading Enzyme in Spinach as an Indicator Enzyme. A number of authors have studied the mechanisms for the formation of volatiles as the results of oxidation of lipids in vegetables (*20, 21*). The enzymes involved are lipases, phospholipid degrading enzymes, galactolipid degrading enzymes, and lipoxygenase. Lipids are hydrolyzed to give free fatty acids and the resulting unsaturated fatty acids are likely to be oxidized to form lipid peroxides by the action of an enzyme, iron-sulfur center, autoxidation, or photoxygenation. In spinach leaves, unsaturated fatty acids (18 : 3) are present in a large portion, which is greater than 73-87 % (*4*). Thus, those free fatty acids would be very rapidly oxidized. The peroxides are then cleaved by a specific enzyme or a chemical reaction. The common $C_3-C_{10}$ aldehydes, ketones, and alcohols, derived from oxidative unsaturated fatty acids breakdown are detected in vegetables and fruits (Figure 7). In spinach leaves, activity of lipoxygenase has not been known (*2*). Therefore, galactolipid and/or phospholipid degrading enzyme(s) may be mainly responsible for providing initial free fatty acids which could be oxidized to form peroxides by either chlorophyll sensitized photoxidation or iron-sulfur center. Free fatty acids were most likely to be produced from phospholipids via an alternative pathway. First, phosphatidyl choline would be hydrolyzed to diglyceride and phosphorylcholine, and the resulting diglyceride is converted to galactolipid in the presence of galactose in spinach leaves. The galactolipid produced would be readily hydrolyzed further to free fatty acid by galactolipase.

**Application of Kinetic Data to Thermal Processing.** In most studies on thermal inactivation of indicator enzymes including peroxidase, lipoxygenase, and LAHase, reaction rate constants and thermodynamic parameters have been determined on the assumption that thermal inactivation of the enzymes follows first order reaction kinetics (*22*). However, a deviation from first order kinetics is generally observed from the residual activity curve. This deviation has been explained by several mechanisms, including the formation of enzyme aggregate with different heat stabilities, the presence of heat stable and labile enzymes, and the series type inactivation kinetics.

Chang et al. (*23*) determined that the apparent reaction order for thermal inactivation of horseradish peroxidase was 1.5. Park et al. (*24*) investigated thermal inactivation for potato lipoxygenase and found that a reaction order of 2.0 could be used to model lipoxygenase inactivation for thermal processing adequately. Lipid acyl hydrolase, including phospholipase C and D, galactolipid degrading enzymes, also showed deviation from first order kinetics as shown in Figure 3 and 5. Thus, thermal processes evaluated with first order inactivation kinetics could result in significant amount of residual activities of the enzymes which might cause quality deterioration of foods. The reaction orders of phospholipase C and D are under investigation in our laboratory.

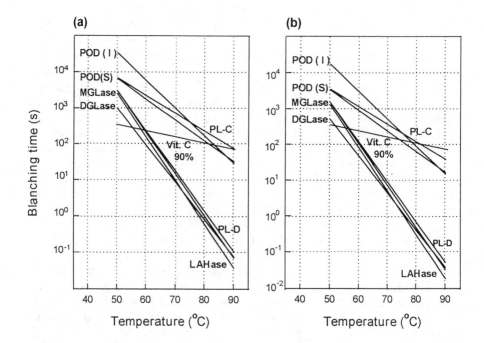

Figure 8.   The nomographs for the determination of 1 % (a) or 10 % (b) residual activities of various enzymes.   POD (I) represents isolated peroxidase; POD (S), peroxidase in suspension; MGLase, monoglyceride lipase; DGLase, diglyceride lipase; PL-C, phospholipase C; PL-D, phospholipase D; LCase, lecithinase; and Vit. C 90 %, vitamine C as a control for the retention of 90 % active form.

The plots shown in Figure 8 were established on the assumption of a first order kinetic model. It shows the nomographs for the determination of 1 or 10 % residual activities of enzymes under the conditions of various temperature-time combinations, indicating the heat treatment time required for adequate blanching (25).

**Application of Phospholipase for a Time-Temperature Indicator (TTI).** Various lipid hydrolyzing enzymes such as lipase could be used for TTIs (26), but stable substrate emulsion is necessary to produce a reproducible reaction rate. The emulsion of triglyceride, a substrate for lipase, may be unstable. It is practically impossible to obtain uniform droplet size in the emulsion and it cannot be stored over very long periods, especially at low temperatures. Therefore, the device using an unstable emulsion system is not reliable to predict enzymatic reactions to an acceptable accuracy and reproducibility. Phospholipid, a substrate of phospholipase, with both hydrophilic and lipophilic portions in the molecule maintains a stable emulsion system. Our objectives were to develop a full-history TTI using a phospholipid-phospholipase system to improve the stability of the substrate emulsion of a TTI at sub-zero termperatures, thereby making it possible to monitor quality changes of frozen foods during storage.

Fifteen percents (v/v) glycerol was added to the mixture of substrate, pH-indicators, and emulsifier in buffer solution. The TTI reaction mixture was not frozen at temperatures as low as -30 ℃. The distinctive color change was detected when bromothymol blue, neutral red, and methyl red were mixed at the ratio of 8:2:0.1. The reaction constant was determined for the temperatures between 30 ℃ and -18 ℃. Standard deviations of the initial rate constants for phospholipase were less than those of lipase at all reaction temperatures (Table V). The relationship between

Table V. The Reaction Rate Constants of the Lipid Hydrolysis Reactions by the Two Lipases at Various Temperatures

| Temp. | Phospholipase | | Lipase | |
|---|---|---|---|---|
| (℃) | $k_a$ (1/min) | $SD^a$ | $k_a$ (1/min) | $SD^a$ |
| 30 | 0.0012919 | 0.0001459 | 0.00195278 | 0.000228 |
| 20 | 0.0006749 | 7.942E-05 | 0.00134049 | 0.000282 |
| 10 | 0.0002048 | 3.875E-05 | 0.00079098 | 0.000210 |
| 0 | 3.706E-05 | 8.561E-06 | 0.00030100 | 0.000116 |
| -5 | 1.498E-06 | 4.081E-06 | 0.00024044 | 5.79E-05 |
| -10 | 2.416E-06 | 4.528E-07 | 2.3691E-05 | 6.58E-06 |
| -15 | 8.772E-07 | 1.769E-07 | 7.6187E-07 | 1.56E-07 |
| -18 | 5.524E-08 | 1.779E-08 | 9.5927E-08 | 1.20E-08 |

Source : reprinted from rf. 27, a) SD : standard deviation

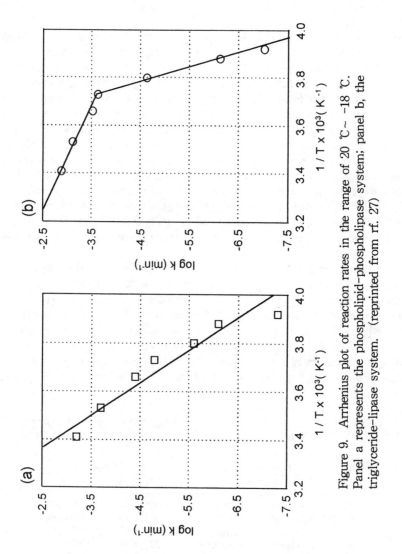

Figure 9. Arrhenius plot of reaction rates in the range of 20 ℃~ –18 ℃. Panel a represents the phospholipid-phospholipase system; panel b, the triglyceride–lipase system. (reprinted from rf. 27)

reaction rate constant and temperature was interpreted by Arrehnius plot. The plot of the lipase reaction was not a straight line (Figure 9b). The curve was very steep below 0 ℃, probably due to the increase of viscosity of the reaction solution. The calculated activation energy of lipase was 11.1 ±0.8 kcal/mole above -5 ℃ and 82.7±3.9 kcal/mole below -5 ℃. The reasonable activation energy of the TTI should be in the range of 25-30 kcal/mole (28). The activation energy of the TTI below -5 ℃ seemed to be too large to monitor food deterioration. The Arrhenius relationship of the phospholipase reaction and temperature was a straight line with a good linear fit (Figure 9a). The activation energy of the phospholipase reaction was 32.1±3.4 kcal/mol. The results thus indicated that the reproducibility of the reaction system of phospholipase was better than that of lipase and that it is more suitable to employ the phospholipase system. Moreover, the reaction system of phospholipase or LAHase from plant tissues such as spinach leaves would provide more favorable properties for monitoring quality change of vegetables during storage at very low temperatures and development of the system is now in progress.

## Conclusions

Thermal inactivation of enzymes in vegetables and fruits can be determined better by in-situ method which is close to natural environment. Lipid-acyl hydrolase such as phospholipid and galactolipid degrading enzymes would be suitable as an indicator enzyme for characterizing blanching process of spinach and other vegetables. Phospholipase C and D were detected in spinach leaves and might play an important role for quality deterioration of vegetables during storage. The phospholipid-phospholipase system provided more reliable kinetic data for a time-temperature indicating device than the triglyceride-lipase system, and is more suitable for predicting quality changes of food.

## Acknowledgements

We thank Prof. A. Fricker and Dr. R. Duden at Federal Research Center for Nutrition, and Prof. M. Loncin at University of Karlsruhe, Germany for their thoughtful discussion on the study. Many thanks are due to T. J. Kim, B. C. Mihn, and M. J. Lee for their assistance in preparing the manuscript. Part of this study was supported by Research Center for New Bio-Materials in Agriculture, Korea.

## Literature cited

1.  Barrett, D. M. ; Theerakulkait, C.  *Food Technol.* **1995**, *49*, 62-65.
2.  Williams, D. C. ; Lim, M. H. ; Chen, A. O. ; Pangborn, R. M. ; Whitaker, J. R.  *Food Technol.* **1986**, *40*, 130-140.

3.  Hirayama, O. ; Oida, H.  *J. Agr. Chem. Soc. Jpn.* **1969**, *43*, 423-428.
4.  Heemskerk, J. W. ; Bögemann, G. ; Helsper, J. P. F. G. ; Wintermans, J. F. G. M.  *Pland Physiol.* **1988**, *86*, 971-977.
5.  Lee, F. A. ; Mattick, L. R.  *J. Food Sci.* **1961**, *26*, 273-275.
6.  Pendlington, S.  *1st Intern. Congr. Food Sci. and Technol. Abstr. s. 120* **1962**.
7.  Lea, C. H. ; Parr, L. J.  *J. Sci. Food Agr.* **1961**, *12*, 785-790.
8.  Oursel, A. ; Grenier, G. ; Tremolieres, A.  *In Proc. 9th Int. Sym. Plant lipids, Kent.* Eds.; Quinn, P. J. and Harwood, J. L. **1990**, pp. 316-318.
9.  List, G. R. ; Mounts, T. L. ; Lanser, A. C. ; Holloway, R. K.  *J. Am. Oil Chem. Soc.* **1990**, *67*, 867-871.
10. Kuroshima, T. ; Hayano, K.  *Soil Sci. Plant Nutr.* **1982**, *28*, 535-542.
11. Lee, H. ; Choi, M. U. : Koh, E. H.  *Korean Biochem. J.* **1989**, *22*, 487-493.
12. Park, K. H. ; Fricker, A.  *Z. Ernährungswiss.* **1977**, *16*, 81-91.
13. Irvine, R. F. ; Letcher, A. J. ; Meade, C. J. ; Dawson, R. M.  *J. Pharmacol. Methods* **1984**, *12*, 171-182.
14. Sahsah, Y. ; Phamthi, A. T. ; Macauley, H. R. ; Arcy-Lameta, A. A. ; Repellin, A. ; Suily-Fodil, Y.  *Biochim. Biophys. Acta.* **1994**, 66-73.
15. Oursel, A. ; Escoffier, A. ; Kader, J. C. ; Dubacq, J. P. ; Trémolieres, A.  *FEBS Letters,* **1987**, *219*, 393-399.
16. Park, K. H. ; Loncin, M. ; Fricker, A.  *Z. Ernährungswiss.* **1977**, *16*, 98-106.
17. Park, K. H. ; Kim, Z. U. ; Shin, J. D. ; Noh, B. S.  *Korean J. Food Sci. Technol.* **1979**, *11*, 171-175.
18. Yoon, J. R. ; Park, K. H.  *Korean J. Food Sci. Technol.* **1982**, *14*, 125-129.
19. Jee, W. J. ; Chon, N. S. ; Kim, I. C. ; Park, K. H. ; Choi, E. H.  *Korean J. Food Sci. Technol.* **1991**, *23*, 442-446.
20. Buttery, R. G.  *In Flavor Chemistry of Lipid Foods.* Ed.; Min, D. B. ; AOCS, **1989** Chapter 8.
21. Furuta, S. ; Nishiba, Y. ; Suda, I.  *Biosci. Biotech. Biochem.* **1995**, *59*, 111-112.
22. Park, K. H. ; Duden, R. ; Fricker, A.  *Z. Ernährungswiss.* **1977**, *16*, 107-114.
23. Chang, B. S. ; Park, K. H. ; Lund, D. B.  *J. Food Sci.* **1988**, *53*, 920-923.
24. Park, K. H. ; Kim, Y. M. ; Lee, C. W.  *J. Agric. Food Chem.* **1988**, *36*, 1012-1015.
25. Paulus, K.  *J. Food Sci.* **1979**, *44*, 1169-1172.
26. Taoukis, P. S. ; Fu, B. ; Labuza, T. P.  *Food Technol,* **1991**, *45*, 70-82.
27. Yoon, S. H. ; Lee, C. H. ; Kim, D. Y. ; Kim, J. W. ; Park, K. H.  *J. Food Sci.* **1994**, *59*, 490-493.
28. Taoukis, P. S. ; Fu, B. ; Labuza, T. P.  *J. Food Sci.* **1989**, *54*, 783-788.

# Chapter 16

# Rapid Test for Alcohol Dehydrogenase During Peanut Maturation and Curing

Si-Yin Chung[1], John R. Vercellotti[1], and Timothy H. Sanders[2]

[1]Agricultural Research Service, U.S. Department of Agriculture,
Southern Regional Research Center, P.O. Box 19687,
New Orleans, LA 70179-0687
[2]Agricultural Research Service, U.S. Department of Agriculture,
Marketing Quality Handling Research, North Carolina State University,
Raleigh, NC 27695-7624

A rapid test for detection of alcohol dehydrogenase (ADH) in peanuts was developed to examine ADH activity during peanut maturation and curing. The test, a microtiter assay, utilized such reagents as ethanol, $NAD^+$, aldehyde dehydrogenase (ALDH), diaphorase, and INT-violet dye. Using the test, a direct relationship between ADH activity and peanut maturity (as defined by progressive peanut pod mesocarp color yellow, orange, brown, and black) was found. The ratio of mature-to-immature (i.e., black-to-yellow) ADH activity in uncured peanuts was 9:1 as compared to 2:1 for other enzymes such as aminopeptidase and peroxidase. The ADH activity of all maturity classes increased to similar levels (i.e., approximately 1:1 ratio) after curing to reduce moisture content to approximately 10%. The finding indicates that ADH activity increased with peanut maturity and increased substantially during peanut curing. The possible mechanism of ADH increase was discussed.

Maturity of peanuts has been shown to be important to many quality factors including flavor and shelf life (*1, 2*). Because of the indeterminate flowering pattern of peanuts, at harvest pods of various maturity are found on the plants. Research has been conducted to define many of the oil, carbohydrate and protein physiological changes that take place during maturation (*3-6*). Arginine (*7*), proteins (*6*) and peptides (*8*) have been related to maturity in peanuts. Williams and Drexler (*9*) found that pod mesocarp color changes in a defined manner with maturation.

Preliminary studies in this laboratory (*10*) showed that the enzyme alcohol dehydrogenase (ADH) appeared to increase with peanut maturity. Pattee et al. (*11*) reported some similarity in ADH profile. However, their work showed that ADH activity increased in very early maturity stages and then varied at the maximum level for the remaining stages. This difference between our work and Pattee et al. (*11*) could be due to sample handling and the method by which the enzyme was prepared. In the work of Pattee et al. (*11*), the enzyme was prepared as a crude extract according to previous work (*12*). It is, therefore, the objective of this study to further define the ADH-maturation relationship and develop a rapid test for ADH. The effect of curing on ADH activity relative to peanut maturity was also investigated.

## Materials and Methods

**Materials.** Ethanol, nicotinamide adenine dinucleotide, aldehyde dehydrogenase (100 units), p-iodonitrotetrazolium violet (INT-violet), diaphorase (20 units), o-phenylenediamine dihydrochloride, and leucine-nitroanilide were purchased from Sigma Co. (St. Louis, MO). Microtiter plates of Immulon I and a microtiter plate reader (Model M700) were purchased from Dynatech Laboratories, Inc. (Chantilly, VA). Bicinchoninic Acid (BCA)-Protein Assay Kit was purchased from Pierce Chemical Co. (Rockford, IL). Peanuts (*Arachis hypogaea* L., var. Florunner; crop year 1990 and 1991) were planted at the USDA-ARS National Peanut Research Laboratory (Dawson, GA), dug 120 days after planting and subjected to windrow drying where samples were taken 0, 1, 2, 3, and 4 days after windrow drying, and day-4 samples were further dried to 10% moisture content with heated air and used for ADH activity comparison among maturity classes after curing. Day-0 samples were used for ADH activity comparison among maturity classes before curing. At each sample date, peanuts were subjected to gentle abrasion to remove the exocarp, sorted by pod color, handed shelled, and seeds were stored at -80°C. Peanut maturity (defined as yellow, orange, brown, and black) was based on the visual hull-scrape/color method (*9*).

**Preparation of Peanut Extracts.** Extracts were prepared by a modification of the method of Chung et al. (*8*). Prior to extraction, seed coat was removed from peanuts of different maturity and curing stages, and peanuts defatted by grinding with a Wiley-mill in sequence with cold acetone and hexane. After air-drying, the resultant defatted peanut meals were stored at -20 °C or used for preparation of crude enzyme extracts. Extracts were prepared by suspending 0.2 g defatted peanut meal in 0.02 M sodium phosphate buffer, pH 8 (1.5 mL) and stirring for 30 min at 4°C. Crude cell extracts were centrifuged at 10,000 rpm for 10 min. Ammonium sulfate was added to 40% saturation [24.2 mg $(NH_4)_2SO_4$ per 100 $\mu$L], and the mixture was incubated for 20 min at 4°C (vortexed every 5 min), and centrifuged. The final supernatant was used for protein determination and assay of alcohol dehydrogenase.

**Rapid Test for Detection of Alcohol Dehydrogenase (ADH) in Peanuts.** The test was a modified version of the aldehyde dehydrogenase (ALDH)-amplified microassay developed by Chung et al. (*13*). Prior to the assay, volumes of extracts from peanuts of different maturity (i.e., yellow, orange, brown, or black) were adjusted so that all of them had the same protein concentration. Protein concentration was determined using the BCA-Protein Assay Kit from Pierce. Briefly, a peanut extract (5 $\mu$L; contained 45 $\mu$g proteins) was added to a mixture containing 0.02 M sodium phosphate buffer, pH 8 (99 $\mu$L), ethanol (3 $\mu$L), diaphorase (3 $\mu$L), and 1.98 mM INT-violet (28 $\mu$L; allowed to warm 50°C to and cool before use). This was followed by addition of ALDH (2 $\mu$L; 10 mg/mL) and 14.3 $\mu$M NAD$^+$ (10 $\mu$L). The whole mixture was then transferred to the well of a microtiter plate. The color developing in the well (i.e., absorbance) was read at 10 min against a blank (contained all reagents except the peanut extract) at 570 nm using a microtiter plate reader. The total volume for the assay was 150 $\mu$L. ADH activity was defined as $A_{570}$/min/mg protein and the mean (n= 3) was obtained from three separate determinations.

**Assay of Aminopeptidase and Peroxidase in Peanuts**.  Aminopeptidase and peroxidase were assayed in a microtiter plate format using the substrates leucine-nitroanilide and *o*-phenylenediamine dihydrochloride (OPD), respectively. Briefly, for assay of aminopeptidase, 50 $\mu$L of leucine-nitroanilide (10 mM) was added to the well of a microtiter plate containing 100 $\mu$L of a diluted peanut sample in 0.05M Tris-HCl, pH 8.6 with 5 mM MgCl$_2$. After 15 min, the color reaction was stopped by adding 50 $\mu$L of 3 M acetate buffer, pH 4.2.   The absorbance was read at 410 nm. A blank containing all reagents except the peanut sample was also prepared. Enzyme activity was expressed as $A_{410}$/min/mg protein.   For assay of peroxidase, a similar protocol was used except that the substrate was OPD (1 mg/mL) in 0.1 M citrate-phosphate buffer, pH 5 with 0.03% hydrogen peroxide, and the absorbance was read at 490 nm after termination of the color reaction with 50 $\mu$L of 4 N sulfuric acid.   Enzyme activity was expressed as $A_{490}$/min/mg protein.

**Results and Discussions**

**Rapid Test for ADH**.  In the rapid test, aldehyde dehydrogenase (ALDH) was used as an amplifier (*13*) to enhance the assay sensitivity.   In the event that ALDH is not available, ADH from a peanut sample can also be detected using the non-amplified assay (*13*). In this case, the volume of the peanut sample to be assayed has to be increased (4 x volume used in an amplified assay) because the non-amplified assay is 4-fold less sensitive than the amplified assay.

**Relationship Between ADH and Maturity**.  Figure 1 shows the typical ADH activity displayed by extracts from mature ("black") and immature ("yellow") peanuts (crop year 1991) in the microtiter plate assay.   As shown, the ADH activity displayed by mature peanuts was approximately 9 times faster than that displayed by immature peanuts. The ADH activity was much higher in mature peanuts, and the mature-to-immature ADH activity ratio was 9:1 (Table I). Previously, Pattee et al. (*11*) showed that after increasing in early maturity stages ADH fluctuated in activity during further peanut maturation.   However, in this study, ADH activity as reflected by the colorimetric assay increased in the following maturity order: yellow< orange< brown< black (Figure 2, lower curve).

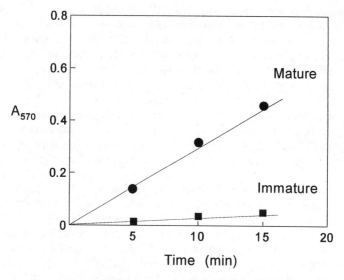

**Figure 1.** Detection of ADH in mature and immature peanuts by the rapid test. All assays were based on the same protein concentration. ADH activity is expressed as $A_{570}$/min/mg protein. Mature and immature refer to seeds from "black" and "yellow" peanut pods.

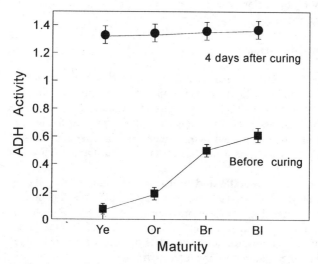

**Figure 2.** Relationship between ADH activity ($A_{570}$/min/mg protein) and peanut maturity before and 4 days after curing. Ye = yellow; Or = orange; Br = brown; and Bl = black.

To further confirm our finding, peanuts from a different crop year 1990 were also investigested.  For crop year 1990, the profile of ADH activity relative to maturity was found to be very similar to that of crop year 1991 in Figure 2, and the ratio of mature-to-immature ADH activity was 7:1 (Table I).  This further suggests that there is a positive relationship between ADH and maturity.  Data presented by Pattee et al. (*11*) revealed a mature (stages)-to-immature (stages) ADH activity ratio of approximately 7.  For comparison, other enzymes such as aminopeptidase and peroxidase were also tested in this study.  These enzymes were selected because like ADH they are abundant in peanuts with a potential influence on flavors and can be detected colorimetrically using substrates such as leucine-nitroanilide and o-phenylenediamine dihydrochloride for aminopeptidase and peroxidase, respectively.  However, in neither case were these enzymes shown to give activity significantly higher in mature peanuts than in immature.  Table I shows the ratios of mature-to-immature activity of these enzymes for crop years 1990 and 1991.  In both cases, the ratios were very similar, namely, 2:1 for aminopeptidase and 1:1 for peroxidase.  These ratios were low as compared to the ADH ratio (i.e., 9:1) (Table I), suggesting that there was little change in activities of aminopeptidase and peroxidase during peanut maturation.

**Table I.  Identification of ADH as Biochemical Marker by Ratio of Enzyme Activity**

| Enzyme[1] | Ratio of Activity (mature-to-immature)[2] | |
|---|---|---|
| | Crop year 1991 | Crop year 1990 |
| Alcohol dehydrogenase (ADH) | 9:1 | 7:1 |
| Aminopeptidase | 2:1 | 1:1 |
| Peroxidase | 1:1 | 1:1 |

[1] Prepared from a crude peanut extract treated with ammonium sulfate (40% saturation).  All enzyme assays (mean activity, n = 3) were colorimetric.
[2] Mature and immature refer to seeds from "black" and "yellow" peanut pods.

The marked increase of ADH activity during peanut maturation is not a unique case.  There have been reports on the increase of ADH during ripening or growth in banana (*14*), tomato (*15, 16*) and other plant tissues (*17*).  The increase of ADH is thought to be a response to and indication of continuing metabolism of carbohydrates under anaerobic conditions (*18-20*).  Under those conditions, increased amount of glycolytic enzymes, ADH and pyruvate decarboxylase were detected.  The latter enzyme participates in the reaction where pyruvate (the intermediate of glycolysis) is converted to acetaldehyde, and acetaldehyde in turn is converted to ethanol by ADH.  It is thought that the metabolic pathways could be the ADH-inducing pathway (*18*).  Because Vercellotti et al. (*4*) showed that during peanut maturation carbohydrate metabolism continued and that the carbohydrate content decreased with peanut maturity, we believe that the increase of ADH found in this study is a response to the continuing metabolism of carbohydrates. In this case, the lower the carbohydrate content, the higher the ADH activity in mature peanuts. Another possible explanation for the increase of ADH is water stress. As there is a reduction in moisture content during peanut maturation, a water stress is imposed. Water stress has been shown to be capable of inducing anaerobiosis (21, 22) which in turn may lead to the metabolism of carbohydrates in peanuts, and consequently the resulting increase of ADH found in this study. The effect of stress on ADH will be discussed below.

**Effect of Curing on ADH.** Figure 3 shows the changes and difference between mature and immature peanuts in ADH activity due to curing. Both groups of peanuts exhibited an increase in ADH activity during curing. The increase was more rapid and pronounced in immature peanuts than in mature; as a result, ADH activity in immature peanuts was almost the same as that in mature at the final curing stage (i.e., Day 4). Similar profiles were also obtained with seed from "orange" and "brown" peanut pods, and subsequently all the maturity classes had similar ADH activity at the final stage of curing (Figure 2, upper curve). This finding suggests that windrow drying had a greater effect on peanuts from early maturity stages in the aspect of ADH activity, and that several factors (e.g., moisture loss and environment) might contribute to the increase of ADH in both mature and immature peanuts during curing. Environment is a possible factor based on the observation (data not shown) that ADH activity after stackpole curing was much lower as compared to that in windrow drying. The stackpole method was used years ago to cure peanuts in the field and was accomplished by stacking peanut plants 6-8 feet high around a pole. The cause for the lower ADH activity in stackpole curing is possibly due to the fact that peanuts were exposed to a different set of environment parameters as compared to peanuts in windrow drying (23).

As indicated above, curing leads to the increase of ADH. The increase of ADH is potentially a response to the stress conditions such as moisture loss occurring during curing because other stress conditions such as oxygen (15, 20, 24, 25) and acid (16, 26, 27) stresses have been shown to induce high level of ADH. Russell et al. (20) and Newman and VanToai (25) showed that anoxic treatment of soybean induced the transcription and accumulation of ADH mRNA, selective synthesis of ADH protein, and accumulation of ADH activity. A similar induction of ADH due to stress conditions has also been reported in maize (28), rice (24), tomato (15, 16), and pea (29). In one study (30), ADH is indicated as a marker of stress conditions that affect the growth of callus tissue cultures. In addition to ADH, stress conditions are known to induce other enzymes related to glycolysis such as sucrose synthase (31) and pyruvate decarboxylase (18, 19). This suggests that glycolysis or carbohydrate metabolism continues under stress conditions. Vercellotti et al. (4) showed that the carbohydrate content in peanuts decreased during curing (a stress condition in this case). On the basis of this, the stress of curing, carbohydrate metabolism and the induction of ADH are related, and the order of occurrence is stress condition, followed by carbohydrate metabolism and then ADH induction. Carbohydrate metabolism (i.e., formation of pyruvate, acetaldehyde, and then ethanol) is second because it is considered a primary pathway for inducing ADH (*18, 20*). In this case, the increase of ADH found in this study could be a response to carbohydrate metabolism induced by stress in peanut curing, during which the carbohydrate content declines and ADH activity increases.

**Precision and Factor Affecting the Assay.** The precision of the assay is shown in Figure 4 with a standard deviation (SD) of 0.01 and relative standard deviation (RSD) of 2.85%. Ammonium sulfate was a factor that could affect the assay sensitivity. Ammonium sulfate was used because peanuts contain large amount of seed storage proteins, the majority of which are globular proteins and have to be removed to enhance assay sensitivity. The removal of these proteins can be done by treatment of the sample with ammonium sulfate at 40% saturation. This treatment not only prevented sample cloudiness that affected the assay, but also improved the mature-to-immature ADH activity ratio, which increased from 3:1 (untreated) to 9:1 (treated). This increase in the ratio was probably due to the difference in the amount of globular proteins removed by ammonium sulfate from mature and immature peanuts. *Note*: higher levels of globular proteins are known to exist in mature peanuts; as more globular proteins were removed from mature peanuts, the specific ADH activity (i.e., $A_{570}$/min/mg protein) increased accordingly because the final total protein level in the sample was reduced. In general, the removal of globular proteins from a crude peanut extract is achieved by treatment with ammonium sulfate at a 40% to 50% saturation (*32, 33*). However, it was noted that ammonium sulfate at a saturation higher than 40% could lower the absorbance, and thus reduce the assay sensitivity. Figure 5 shows the effects of different saturation (40%, 45%, and 50%) of ammonium sulfate on the absorbance. At 45% saturation, the rate was about one third of the control (i.e., 40% saturation), and at 50% saturation, it was almost zero. The reduction in the rate was probably due to the salting-out effect (i.e., precipitation of ADH) or interference of the assay by ammonium sulfate in the sample. *Note*: ammonium sulfate was present in the assay because the sample was not dialysed; dialysis was deemed as not necessary because little difference was found in the absorbance between dialyzed and undialyzed samples at 40% saturation (data not shown).

## Conclusions

We report a rapid test to measure ADH activity in peanuts. We noted that ADH increased during peanut maturation and curing. However, at the final stages of curing, all maturity classes had similar levels of ADH activity. The mechanism for this probably is due to the carbohydrate metabolism occurring during maturation and curing, which potentially induces ADH synthesis. Further investigation is needed to define the role of ADH in peanut maturity.

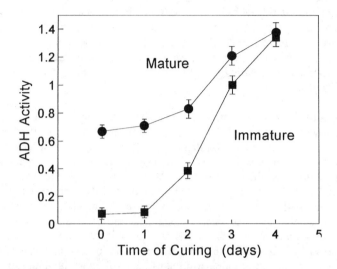

**Figure 3.** Effect of curing on ADH activity ($A_{570}$/min/mg protein) in mature and immature peanuts. Each data point represents the mean of three determinations. Error bars represent standard deviation from the mean.

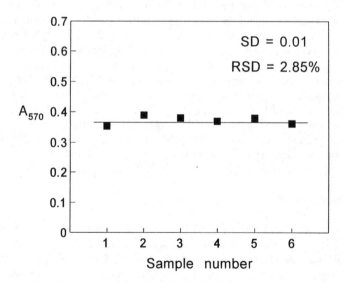

**Figure 4.** Precision of rapid test for ADH. ADH was prepared from uncured mature peanuts.

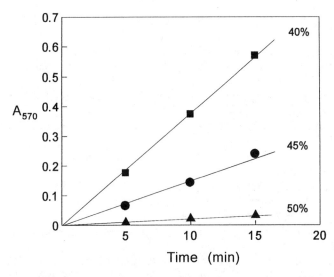

**Figure 5.**   Effect of different saturations of ammonium sulfate on the absorbance. All assays were based on the same protein concentration.

## Acknowledgements

We thank Lisa L. Bothman and Maurice R. Brett for assistance in the preparation of peanut samples. Also, we thank Lisa for assistance in the enzyme assays.

## Literature Cited

1.   Sanders, T.H.; Vercellotti, J.R.; Crippen, K.L.; Civille, G.V. *J. Food Sci.* **1989**, *54*, 475-477.
2.   Bett, K.L.; Boylston; T.D. In *Lipid Oxidation in Food;* St. Angelo, A.J., Ed; ACS Symposium Series 500; Amer. Chem. Soc.: Washington, D.C., 1992; 322-343.
3.   Sanders, T.H.; Lansden, R.L.; Greene, R.L.; Drexler, J.S.; Williams, E.J. *Peanut Sci.* **1982**, *9*, 20-23.
4.   Vercellotti, J. R.; Sanders, T. H.; Chung, S. Y.; Bett, K. L.; Vineyard, B. T.   In *Food Flavors: Generation, Analysis and Process Influence*; Charalambous, G. Ed.; Elsevier Science Publishers: Amsterdam, The Netherlands, 1995, Vol. 34; 1547-1578.
5.   Pattee, H.E.; Johns, E.B.; Singleton; Sanders, T.H. *Peanut Sci.* **1974**, *1*, 57-62.
6.   Basha, S.M. *J. Agric. Food Chem.* **1990**, *38*, 373-376.
7.   Young, C.T.; Mason, M.E. *J. Food Sci.* **1972**, *37*, 722-725.

8. Chung, S.Y.; Ullah, A.H.; Sanders, T.H. *J. Agric. Food Chem.* **1994**, *42*, 623-628.
9. Williams, E.J.; Drexler, J.S. *Peanut Sci.* **1981**, *8*, 134.
10. Chung, S.Y.; Vercellotti, J.R.; Sanders, T.H. 209th ACS National Meeting; Anaheim, CA, April 2-6, 1995: AGFD abstract #50.
11. Pattee, H.E.; Singleton, J.A.; Johns, E.B.; Mullin, B.C. *J. Agric. Food Chem.* **1970**, *18*, 353.-356.
12. Pattee, H.E.; Swaisgood, H.E. Peanut alcohol dehydrogenase. *J. Food Sci.* **1968**, *33*, 250-253.
13. Chung, S.Y.; Vercellotti, J.R.; Sanders, T.H. *J. Agric. Food Chem.* **1995**, *43*, 1545-1548.
14. Tan, S.C.; Ng, A.M.; Wade, N.L. *ASEAN Food J.* **1987**, *3*, 138-143.
15. Bicsak,T.A.; Kann, L.R.; Reiter, A.; Chase, T. Jr. *Arch. Biochem. Biophys.* **1982**, *216*, 605-615.
16. Longhurst, T.J.; Tung, H.F.; Brady, C.J. *J. Food Biochem.* **1990**, *14*, 421-433.
17. McDonald, R.C.; Kimmerer, T.W. *Physiol. Plant.* **1991**, *82*, 582.
18. Tajima, S.; LaRue, T.A. *Plant Physiol.* **1982**, *70*, 388-392.
19. Kimmerer,T.W. *Plant Physiol.* **1987**, *84*, 1210-1213.
20. Russell, D.A.; Wong, D.M.L.; Sachs, M.M. *Plant Physiol.* **1990**, *92*, 401-407.
21. Sprent, J.I.; Gallacher, A. *Soil Biol. Biochem.* **1975**, *8*, 317-320.
22. Heikkila, J.J.; Papp, J.E.T.; Schultz, G.A.; Bewley, J.D. *Plant Physiol.* **1984**, *76*, 270-274.
23. Sanders, T.H.; Lansden, J.A.; Vercellotti, J.R.; Crippen, K.L. *Proc. APRES* **1990**, *22*,71
24. Kadowaki, K.I.; Matsuoka, N.; Murai, N.; Harada, K. *Plant Sci.* **1988**, *54*, 29-36.
25. Newman, K.D.; VanToai, T.T. *Crop Sci.* **1991**, *31*, 1253.
26. Roberts, J.K.M.; Callis, J.; Wemmer, D.; Walbot, V.; Jardetzky, O. *Proc. Natl. Acad. Sci. USA.* **1984**, *81*,3379-3383.
27. Kader, A.A. *Food Technol.* **1986**, *40*, 99-104.
28. Sachs, M.M.; Freeling, M.; Okimoto, R. *Cell.* **1980**, *20*, 761-767.
29. Llewellyn, D.J.; Finnegan, E.F.; Ellis, J.G.; Dennis, E.S.; Peacock, W.J. *J. Mol. Biol.* **1987**, *195*, 115-123.
30. Mangolin, C.A.; Prioli, A.J.; Machado, F.P.S. *Biochem. Genet.* **1994**, *32*, 191-200.
31. Springer, B.; Werr, W.; Starlinger, P.; Bennett, D.C.; Zokolica, M.; Freeling, M. *Mol. Gen. Genet.* **1986**, *205*, 461-468.
32. Neucere, N.J. *Anal. Biochem.* **1969**, *27*, 15-24.
33. Jacks, T.J.; Hensarling, T.P.; Neucere, N.J.; Graves, E.E. *J. Agric. Food Chem.* **1983**, *31*, 10-14.

Chapter 17

# Glycoalkaloids in Fresh and Processed Potatoes

Mendel Friedman and Gary M. McDonald

Agricultural Research Service, U.S. Department of Agriculture, Western Regional Research Center, 800 Buchanan Street, Albany, CA 94710

As part of a program designed to improve food safety by controlling the biosynthesis of glycoalkaloids in potatoes, we define conditions of sampling, handling, storing, shipping, and processing that influence the biosynthesis of potentially toxic glycoalkaloids in potatoes after harvest. This brief overview also suggests research needs to develop a protocol that can be adopted by potato producers and processors to minimize post-harvest synthesis of glycoalkaloids in potatoes. Reducing glycoalkaloid concentration in potatoes will provide a variety of benefits extending from the farm to processing, shipping, marketing, and consumption of potatoes and potato products. Minimizing pre- and postharvest glycoalkaloid production in potatoes, including new cultivars, requires an integrated multi-disciplinary approach.

Steroidal glycoalkaloids have been found in potatoes (1), green tomatoes (2), and eggplants (3). Symptoms of glycoalkaloid toxicity experienced by animals and humans include colic pain in the abdomen and stomach, gastroenteritis, diarrhea, vomiting, burning sensation about the lips and mouth, hot skin, fever, rapid pulse, and headache (4). The reported toxicity of these glycoalkaloids may be due to such adverse effects as: (1) anticholinesterase effects on the central nervous system (5); (2) induction of hepatic ornithine decarboxylase, a cell proliferation marker enzyme (6), and (3) disruption of cell membranes affecting the digestive system (7-9). Toxicity does not seem to occur at the genetic level (10). One manifestation of these adverse effects may be alkaloid-induced teratogenicity (11,12).

The estimated highest safe level of total glycoalkaloids for human consumption is about 1 mg/kg body weight, a level that may cause gastrointestinal irritation (13). The acute toxic dose is estimated to be about 1.75 mg/kg body weight (4). A lethal dose may be as low as 3-6 mg/kg body weight (14).

With respect to potential toxicity of glycoalkaloids to humans, Hopkins (15) points out that although glycoalkaloids are present in foods such as potatoes consumed daily by animals and humans, their toxicological status is poorly defined. Moreover, Rayburn et al. (16-18) point out that depending on the variety, potatoes may contain the glycoalkaloids α-chaconine and α-solanine at concentration ratios of 74:26 to 40:60 (α-chaconine: α-solanine). Therefore, additional studies are needed to address the interactions of glycoalkaloids when consumed as mixtures of varying proportions.

Human consumption of potatoes, an excellent source of carbohydrates and good-quality protein (19), varies by country. For example, the average daily per capita intake in the United States is about 167 g (20); in the United Kingdom, 140 g (15); and in Sweden, 300 g (13). The cited amount for the United Kingdom is estimated to contain 14 mg of glycoalkaloids. Although the glycoalkaloid content of most commercial potato varieties is usually below a suggested guideline of 200 mg/kg fresh potatoes (21), the content can increase significantly on exposure to light and as a result of mechanical injury, including peeling and slicing (22-23). After processing, further accumulation of glycoalkaloids is halted as the necessary enzymes for biosynthesis have been deactivated. However, since glycoalkaloids are largely unaffected by home processing conditions such as baking, boiling, frying and microwaving (1,2,24-29), any glycoalkaloids present in the tubers before cooking will still remain afterwards. Since the two major potato glycoalkaloids α-chaconine and α-solanine differ in biological potency, may act synergistically, and their ratio may vary in different cultivars (18,30,31), care should be exercised in relating dose-response data for individual glycoalkaloids to the total amount in potatoes. New varieties may also contain glycoalkaloids of unknown structure and function inherited from their progenitors. These considerations suggest a need to reduce steroidal glycoalkaloid levels in the diet, possibly through suppression of enzymes responsible for their biosynthesis in plants (32-34).

For the purposes of this study, we define the following terms: *glycoalkaloids* - naturally occurring, nitrogen-containing plant steroids with a carbohydrate side chain attached to the 3-hydroxy position, e.g. α-chaconine and α-solanine from potatoes, α-tomatine from tomatoes, and solasonine from eggplants; *aglycones* - the steroidal parts of the glycoalkaloid lacking the carbohydrate side chain, e.g. solanidine from α-chaconine and α-solanine, tomatidine from α-tomatine, and solasodine from solasonine; *alkaloids* - glycoalkaloids and aglycones.

Table I summarizes glycoalkaloid content of some potato varieties grown in the United States, Table II lists the content of different varieties grown in Sweden, and Table III shows glycoalkaloid content of widely consumed potato products.

Figure 1 illustrates structures of common and uncommon *Solanum* glycoalkaloids, Figure 2 depicts structures of hydrolysis products of α-chaconine and α-solanine, Figures 3 and 4 show HPLC chromatograms of α-chaconine and α-solanine and hydrolysis products, and Figure 5 correlates HPLC and immunoassay measurements of potato glycoalkaloids.

Table I. α-Chaconine and α-Solanine Content of Various Potato Parts and Cultivars (*I*)

| material | α-chaconine (mg/100g fresh wt.) | α-solanine (mg/100g fresh wt.) | ratio α-chaconine/ α-solanine |
|---|---|---|---|
| sprouts (no. 3194) | 150.4 | 123.4 | 1.22 |
| berries | 22.1 | 15.9 | 1.39 |
| tubers (Lenape) | 13.5 | 5.9 | 2.28 |
| peel (Lenape) | 62.3 | 23.0 | 2.70 |
| flesh (Lenape) | 8.02 | 3.95 | 2.03 |
| tubers (no. 3194) | 3.68 | 1.95 | 1.89 |
| tubers (Simplot I) | 3.85 | 1.72 | 2.24 |
| tubers (Simplot II) | 2.75 | 1.07 | 2.57 |
| tubers (commercial red) | 2.72 | 1.09 | 2.31 |
| tubers (commercial white) | 1.17 | 0.58 | 2.02 |
| tubers (Idaho Russet) | 1.34 | 0.65 | 2.06 |
| tubers (Washington Russet) | 1.30 | 0.58 | 2.24 |

Table II. Glycoalkaloid Content of Commercial Swedish Early Potato Varieties (Adapted from ref. 35)

| variety | total glycoalkaloids (mg/100g fresh wt) | | ratio α-chaconine/α-solanine |
|---|---|---|---|
| | mean + SD | min-max | |
| Ulster Chieftan | 22.1 ± 5.0 | 15.4 - 34.4 | 1.46 |
| Silla | 16.2 ± 2.6 | 11.6 - 21.5 | 1.86 |
| Early Puritan | 14.3 ± 2.4 | 11.0 - 19.9 | 1.63 |
| Maria | 9.8 ± 2.3 | 5.7 - 14.3 | 1.41 |
| Evergood Eldorado | 9.5 ± 2.9 | 4.9 - 13.9 | 1.86 |
| Provita | 8.5 ± 2.2 | 5.8 - 13.9 | 1.77 |

**Table III.** α-Chaconine and α-Solanine of Processed Potato Products (mg/100g of fresh weight) (*1*)

| sample | α-chaconine | α-solanine | ratio α-chaconine/ α-solanine |
|---|---|---|---|
| french fries, freeze-dried | 0.42 | 0.42 | 1.00 |
| wedges, freeze-dried | 2.39 | 2.01 | 1.18 |
| skins, sample A | 3.89 | 1.74 | 2.23 |
| skins, sample B | 4.40 | 2.36 | 1.86 |
| skins, sample C | 11.61 | 7.23 | 1.60 |
| skins, sample D | 11.95 | 8.35 | 1.43 |
| pancake powder, brand A | 2.05 | 2.41 | 0.82 |
| pancake powder, brand B | 2.48 | 1.94 | 1.27 |
| mushed flakes | 3.17 | 3.29 | 0.96 |
| chips, brand A | 1.30 | 1.05 | 1.23 |
| chips, brand B | 3.16 | 1.76 | 1.79 |
| chips, brand C | 5.88 | 5.02 | 1.17 |

The main objective of this paper is to briefly describe the fate of potato glycoalkaloids α-chaconine and α-solanine in potato tubers and processed potatoes after harvest.

**Handling and Sampling of Potatoes to Minimize Adverse Effects**

The biosynthesis of glycoalkaloids in potatoes continues long after harvest. Factors which influence glycoalkaloid formation include light, storage conditions, and mechanical injury. Possible relationships of other post-harvest events, such as blackening, blighting, and browning on glycoalkaloid formation are not well defined. In order to standardize handling and sampling of potatoes to minimize glycoalkaloid formation, it would be helpful to know the biochemical basis for the post-harvest changes which affect quality and safety. The following sections offer a brief review of some of these changes and the research approaches for lessening or eliminating such adverse effects.

**Glycoalkaloids, Flavor, and Taste.** Experiments with human taste panels revealed that potato varieties with glycoalkaloid levels exceeding 14 mg/100 g fresh weight tasted bitter (*38,39*). Those in excess of 22 mg/100 g also induced mild to severe burning sensations in the mouths and throats of panel members. In a related study, Kaaber (*40*) demonstrated that the Norwegian potato variety Kerrs Pink was quite susceptible to greening-related glycoalkaloid synthesis and accompanying increases in bitterness and burning sensations, whereas the Bintje variety was not.

Zitnak and Filadelfi-Keszi (*41*) describe the isolation of the diglycoside ß₂-chaconine, a so-called potato bitterness factor. ß₂-Chaconine and other glycoalkaloid hydrolysis products (Figure 2) are readily formed on exposure of the glycoalkaloids in pure form or in potatoes to acid conditions (*42,43*).

**Effect of Light on Glycoalkaloid Synthesis.** Exposure of post-harvest potato tubers to light, whether incandescent, fluorescent, or natural, can dramatically enhance glycoalkaloid synthesis. For example, exposure of the cultivar Sebago to a 15 watt incandescent lamp for 10 days resulted in an increase in glycoalkaloid content from 4.8 to 19 mg/100 g fresh weight (*44*). Exposure of peeled Russet Burbank potato slices to fluorescent light of 200 foot-candles for 48 hr caused an increase from 0.2 to 7.4 mg/100 g fresh weight.

Exposure of commercial White Rose potatoes to fluorescent light for 20 days induced a time-dependent greening of potato surfaces. In addition to increases in chlorophyll content, chlorogenic acid and glycoalkaloid levels also increased, but no changes were observed in the content of inhibitors of the digestive enzymes trypsin, chymotrypsin, and carboxypeptidase A (*23*).

Generally, the increases in chlorophyll and glycoalkaloid synthesis seem to depend on the wavelengths of the light to which the tubers are exposed (*45*).

Since potato cultivars differ significantly in their ability to produce greening-related glycoalkaloids (*46*), it should be possible to find and use varieties with low rates of post-harvest glycoalkaloid synthesis.

**Mechanical Damage, Pest Resistance, and Glycoalkaloids.** Bruising, cutting, and slicing of potato tubers induces the formation of glycoalkaloids (*47,48*). The response increases with the extent of injury and is cultivar-related (*49*). For example, Fitzpatrick et al. (*50*) showed that the glycoalkaloid content of potato slices increased from 5.5 to 99.4 mg/100 g fresh weight after storage for 4 days. Injury such as slicing before storage induces a burst in glycoalkaloid synthesis. It is therefore evident that storing whole tubers and cutting them just before cooking is much preferred over cutting and then storing them to await cooking.

Fungi such as *Fusarium solani* and *Phoma foveata* damage potatoes by causing storage rot (*51*). Olsson (*49,52,53*) attempted to find out whether differences in initial levels of glycoalkaloids of various cultivars and their differing ability to respond to mechanical damage by increasing glycoalkaloid synthesis influenced their resistance to *Fusarium* and *Phoma* fungi. The extent of damage correlated with original glycoalkaloid content - a genotype with a high initial level of glycoalkaloids resulted in a greater increase than one with a low initial level. Cultivars most susceptible to

GLYCOALKALOIDS:

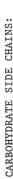

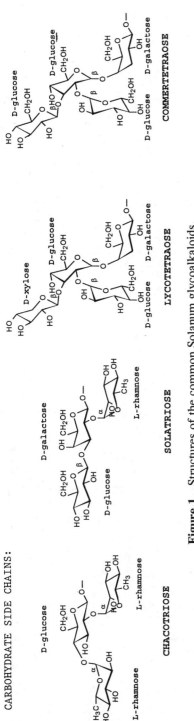

**Figure 1.** Structures of the common Solanum glycoalkaloids.

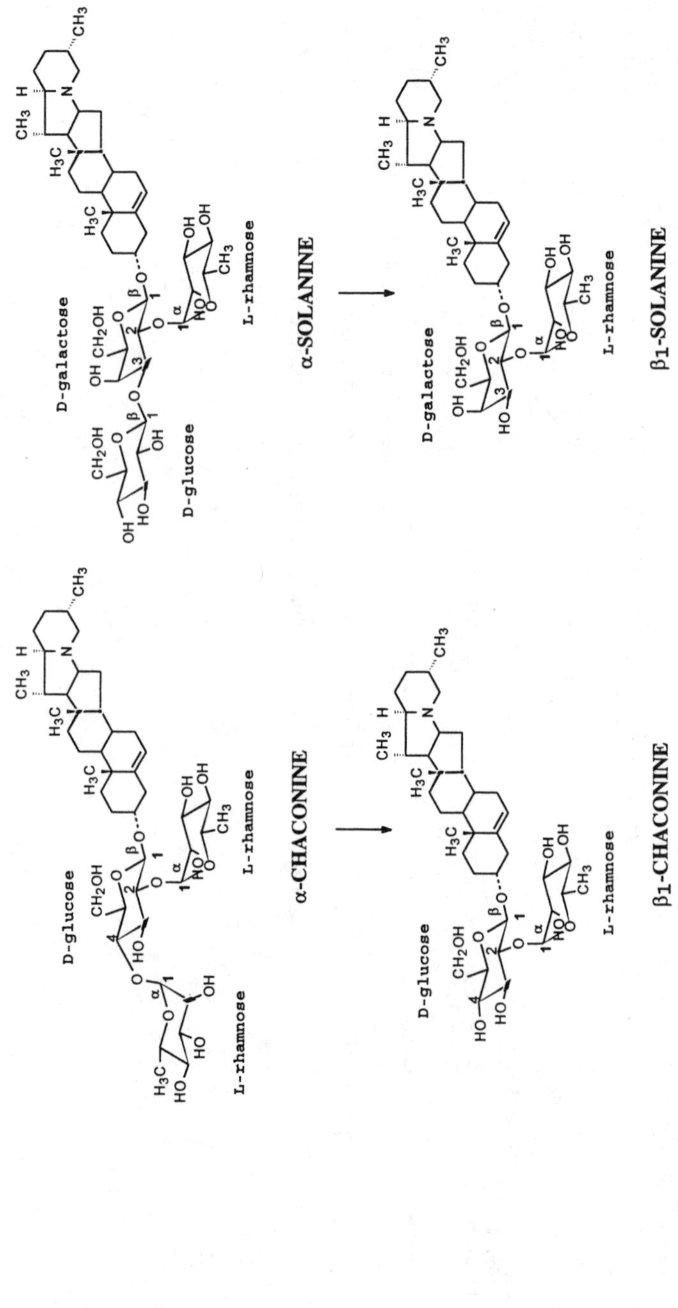

D-galactose
D-glucose
L-rhamnose
α-SOLANINE
D-galactose
L-rhamnose
β1-SOLANINE
D-glucose
L-rhamnose
L-rhamnose
α-CHACONINE
D-glucose
L-rhamnose
β1-CHACONINE

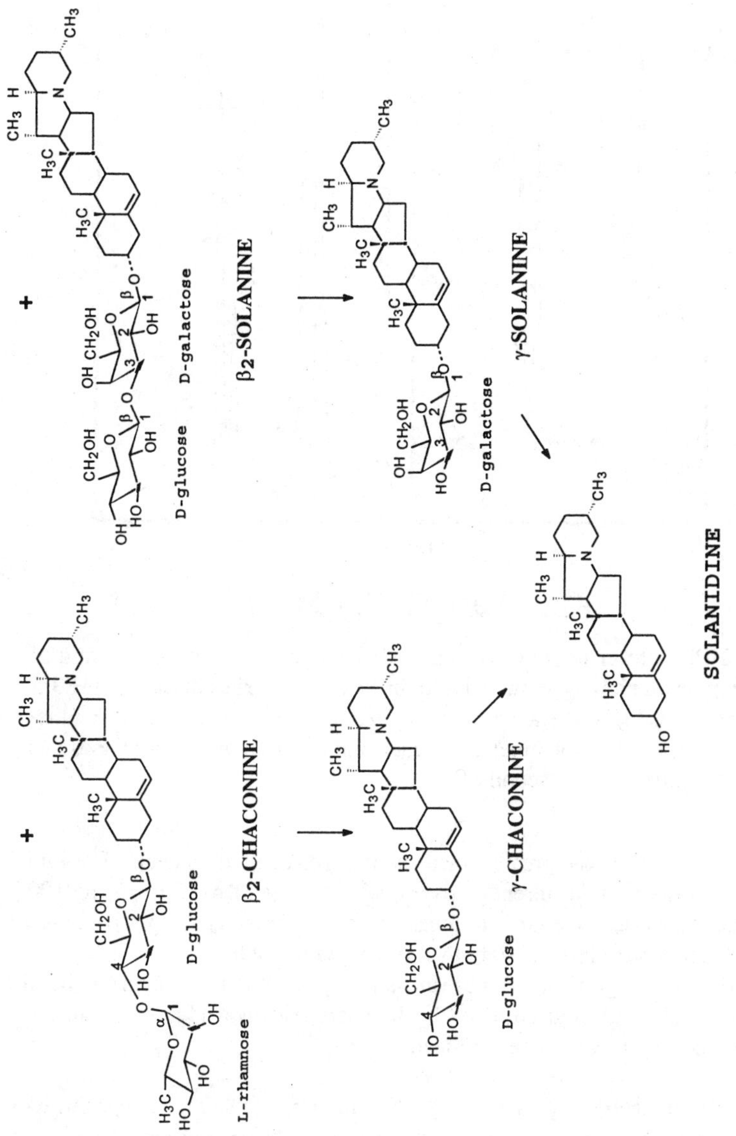

**Figure 2.** Structures of hydrolysis products of α-chaconine and α-solanine.

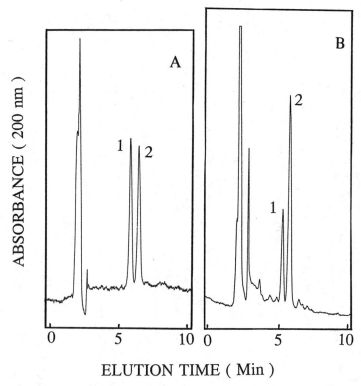

**Figure 3.** HPLC chromatograms (Resolve C18 column; flowrate, 1 mL/min of 100mM ammonium phosphate, monobasic, in 35% acetonitrile adjusted to pH 3.5 with phosphoric acid). (A) standard containing 50 $\mu$g/mL of $\alpha$-solanine and $\alpha$-chaconine. Peaks: 1.$\alpha$-solanine; 2. $\alpha$-chaconine. (B) potato foliar extract. Peaks: 1.$\alpha$-solanine; 2. $\alpha$-chaconine.

mechanical injury showed the greatest increase in glycoalkaloid content. The initial content of glycoalkaloids in different cultivars, which ranged from 2 to 15 mg/100 g fresh weight, did not affect resistance to fungal infection. Some genotypes with higher glycoalkaloid levels were less resistant than others with low levels.

These findings suggest that (a) mechanical injury needs to be minimized; and (b) selection of low-glycoalkaloid cultivars by breeders may not adversely affect the potato's resistance to at least some pathogens.

**Storage Temperature and Humidity**. The influence of storage temperature on glycoalkaloid formation is not clear-cut, possibly because humidities varied widely in many temperature-storage studies (*25*). Immersion of potatoes in water may reduce post-harvest glycoalkaloid formation (*23, 54*). The effects of soaking or spraying with

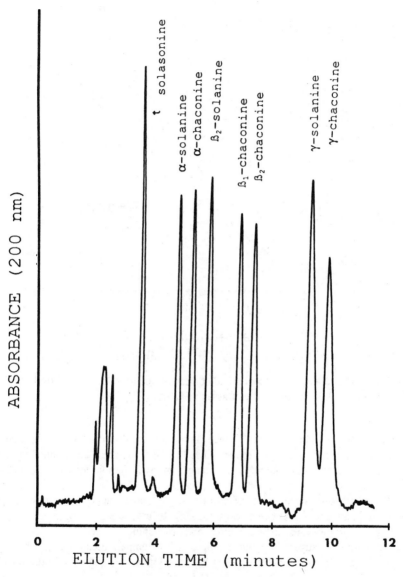

**Figure 4.** HPLC chromatogram (Resolve C18 column; flowrate, 1 mL/min of 100mM ammonium phosphate, monobasic, in 35% acetonitrile adjusted to pH 3.5 with phosphoric acid) of α-solanine and α-chaconine and their respective hydrolysis products (*36*).

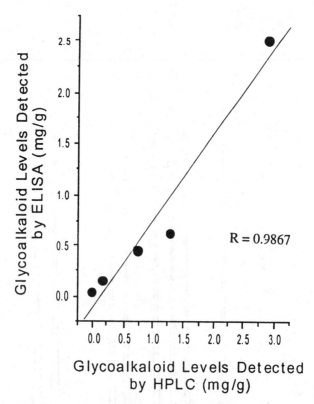

**Figure 5.** Correlation of glycoalkaloid content of potatoes determined by HPLC and enzyme-linked immunosorbent assay (ELISA) (*37*).

water have been little studied. Since humidity could turn out to influence glycoalkaloid formation during storage, a need exists to define the combined effects of storage temperature and humidity on changes in glycoalkaloid levels of potatoes.

**Cold-Induced Sweetening of Potatoes.** Storage temperature may affect potatoes both adversely and beneficially (*55,56*). It influences, among other things, sprouting, respiratory rate, disease control, and sweetening. Sweetening is caused by the accumulation of sugars in potatoes stored at low temperatures. The mechanism of sweetening is being actively explored by Sowokinos (*57*) and Van Berkel et al. (*58*).

Sugars and starch exist together in the tuber and undergo continual enzyme-catalyzed transformations as shown in the following schematic:

$$SUCROSE \leftrightarrow GLUCOSE + FRUCTOSE$$

$$\updownarrow$$

$$STARCH \leftrightarrow GLUCOSE$$

Above 10°C, the sugars and starch remain in balance with the sugars either reforming into starch or being used up in other reactions. Below 10°C, however, reducing sugars start to accumulate in the tuber. This is undesirable because when the tubers are cooked, the sugars can then participate in nonenzymatic Maillard browning reactions causing darkened and off-flavor products. Storage temperatures need to be studied and optimized to minimize sweetening and glycoalkaloid production while maintaining disease resistance.

**Internal Black Spot and Other Discolorations.** Potatoes are susceptible to a variety of discolorations which may adversely affect quality and safety (*59-61*). Internal discoloration can result from browning, whereby the amino acid tyrosine is first hydroxylated to dihydroxyphenylalamine (DOPA) which is then oxidized to dopaquinone. Both events are catalyzed by polyphenol oxidase (PPO). Dopaquinone polymerizes to red and brown pigments in the presence of oxygen. This series of reactions is known as enzymatic browning. Nonenzymatic browning may also occur; for example, amino acids and proteins may polymerize with dopaquinone, or amine groups of amino acids and  proteins may react with reducing sugars (*62-65*). Chlorogenic acid and vitamin C (ascorbic acid) also participate in nonenzymatic browning (*66*). Browning-initiated discolorations may be increased during storage, sorting, packing, and transportation of potatoes.

Chlorogenic acid (3-0-caffeoylquinic acid), which constitutes about 90% of the total polyphenolics in potato tubers, is also responsible for bluish-gray discoloration of boiled and steamed potatoes following exposure to air. This so-called "after cooking-blackening" is perceived by many consumers to be undesirable. The blackening appears to be due to the formation of a reduced ferrous ion-chlorogenic acid complex in the potato. Following exposure to air, the colorless ferrous complex is oxidized to a dark ferric complex. Greening of potatoes is generally accompanied by increased chlorogenic acid content (*23*).

Dao and Friedman (*22*) also evaluated HPLC and UV methods for measuring chlorogenic acid in commercial and experimental potatoes, potato plant parts, cooked potatoes, and commercial potato products. They report chlorogenic values for many samples.

Storage at low temperatures (3.5°C), freezing injuries, heat, and microorganisms also induce internal discoloration (*67*). Friedman and Bautista (*63*), and Molnar-Perl and Friedman (*65*) demonstrated the potential of several structurally different SH containing amino acids and peptides to inhibit browning in fresh and dehydrated potatoes. These and other sulfite-substitutes may also inactivate enzymes catalyzing the biosynthesis of glycoalkaloids. Such inhibition should lead to supprssion of post-harvest synthesis of glycoalkaloids. Ideally, it may be possible to prevent both browning and glycoalkaloid formation with one treatment.

**Inhibition of Potato Sprouting.** Although γ-irradiation and the insecticide chlorfam (CIPC) are good sprouting inhibitors (*68,69*), a need exists to find new inhibitors derived from natural sources. Oxygenated monoterpenes derived from essential oils such as 1,4-cineole and limonene oxide appear to meet this need since they inhibit sprouting and fungal growth under practical conditions of storage (70). To facilitate commercial use of these compounds, further studies are needed on the composition, including glycoalkaloid levels, and nutrition of monoterpene-treated tubers.

Parasitic fungi such as *Phytophthora infestans* induce formation of new compounds, the so-called phytoalexins defined as low-molecular-weight antimicrobial compounds that are both synthesized by and accumulated in plants after their exposure to microorganisms (71). Potato tubers produce about 25 of these stress metabolites, such as rishitinol. Can any of these terpene metabolites be developed into practical protectants against sprouting and infection?

**Future Studies**

The following outlines needed studies to benefit potato growers and consumers.

(1) Determine the *relative susceptibilities* to greening and mechanical damage of the present major commercial varieties and new cultivars and measure accompanying changes in glycoalkaloids, calystegines, chlorogenic acid, tyrosine, and ascorbic acid content. Commercial varieties include the following; Atlantic, Centennial Russet, Katahdin, Kennebec, Norchip, Norgold Russet, Ontario, Russet Burbank, Superior, and White Rose (72).

(2) Evaluate food-compatible enzyme inhibitors, such as citric and sulfur amino acids, and substrate inhibitors for their ability to inactivate SGT and other enzymes which catalyze glycoalkaloid biosynthesis (32).

(3) Evaluate films made from agricultural products with built-in chromophores that absorb light for their ability to protect potatoes against greening, browning, and spoiling (73).

(4) Investigate the effect that size and maturity have on glycoalkaloid levels. As the age of the tuber makes a large difference in concentration of glycoalkaloids (67), care must be taken to insure that in comparative studies, such as determining glycoalkaloid levels of different cultivars, samples consist of tubers of the same level of maturity. Full-grown tubers of approximately the same size appear to be best.

(5) Since many factors, such as light, temperature, and mechanical injury can induce glycoalkaloid production in post-harvest tubers, it is essential to reduce these sources of error when comparing the base levels of different varieties. Samples should be analyzed as soon as possible after harvest. We have had good success immediately flash-freezing and freeze-drying our samples to eliminate any further storage and handling effects. Compare freeze-dried powders versus fresh samples and use the formation to make recommendations for a standard protocol for sampling and handling of potatoes to minimize glycoalkaloid biosynthesis after harvest.

In conclusion, inappropriate post-harvest handling of tubers during storage and shipping increases glycoalkaloid biosynhesis in commercial cultivars. A need exists to devise conditions to minimize this effect during storage, shipping and processing so as to assure that the products reaching the consumer are optimal in appearance, taste, and safety.

## Literature Cited

1. Friedman, M.; Dao, L. *J. Agric. Food Chem.* **1992**, *40*, 419-423.
2. Friedman, M.; Levin, C. E. *J. Agric. Food Chem.* **1995**, *43*, 1507-1511.
3. Aubert, S.; Daunay, M. C.; Pochard, E. *Agronomie* **1989**, *9*, 641-651.
4. van Gelder, W. M. J. In *Poisonous Plants Contaminating Edible Plants*; Rizk, A-F. M., Ed.; CRC: Boca Raton, Florida, 1990; pp 117-156.
5. Roddick, J. G. *Phytochemistry* **1989**, *28*, 2631-2634.
6. Caldwell, K. A., Grosjean, O. K., Henika, P. R.; Friedman, M. *Food and Chem. Toxicol.* **1991**, *29*, 531-535.
7. Blankemeyer, J. T.; Atherton, R.; Friedman, M. *J. Agric. Food Chem.* **1995**, *43*, 636-639.
8. Blankemeyer, J. T.; Stringer, B. K.; Rayburn, J. R.; Bantle, J. A.; Friedman, M. *J. Agric. Food Chem.* **1992**, *40*, 2022-2026.
9. Roddick, J.G.; Rijnenberg, A.L.; Weissenberg, M. *Phytochemistry* **1992**, *31*, 1951-1954.
10. Friedman, M.; Henika, P. R. *Food Chem. Toxicol.* **1992**, *30*, 689-694.
11. Keeler, R. F.; Baker, D. C.; Gaffield, W. In *Handbook of Natural Toxins*; Keeler, R.F.; Wu, A.T., Eds.; Marcel Dekker: New York; 1991, Vol. 6, pp 83-97.
12. Renwick, J. H.; Claringbold, D. B.; Earthy, M. E.; Few, J. D.; McLean A.C.S. *Teratology* **1984**, *30*, 371-381.
13. Slanina, P. *Food Chem. Toxicol.* **1990**, *28*, 759-761.
14. Morris, S. C.; Lee, T. H. *Food Technol. Australia* **1984**, *36*, 118-124.
15. Hopkins, J. *Food Chem. Toxicol.* **1995**, *33*, 323-329.
16. Rayburn, J. R.; Bantle, J. A.; Friedman, M. *J. Agric. Food Chem.* **1994**, *42*, 1511-1515.
17. Rayburn, J. R.; Bantle, J. A.; Qualls, C. W. Jr.; Friedman, M. *Food Chem. Toxicol.* **1995a**, *33*, 1021-1025.
18. Rayburn, J. R.; Friedman, M.; Bantle, J. A. *Food Chem. Toxicol.* **1995b**, *33*, 1013-1019.
19. Markakis, P. In *Protein Nutritional Quality of Foodsand Feeds*; Friedman, M., Ed.; Dekker: New York, **1975**; Part 2, pp 471-487.
20. Friedman, M. *J. Agric Food Chem.* **1996**, *44*, 1-24.
21. Maga, J. A. *Food Reviews International* **1994**, *10*, 385-418.
22. Dao, L; Friedman, M. *J. Agric. Food Chem.* **1992**, *40*, 2152-2156.
23. Dao, L.; Friedman, M. *J. Agric. Food Chem.* **1994**, *42*, 633-639.
24. Bushway, R. J.; Ponnampalam, R. *J. Agric. Food Chem.* **1981**, *29*, 814-817.
25. Chungcharoen, A. *Glycoalkaloid Content of Potatoes Grown Under Controlled Environments and Stability of Glycoalkaloids During Processing.* Ph. D. Thesis, University of Wisconsin, Madison, WI, 1988.
26. Friedman, M. In *Evaluation of Food Safety;* Finley, J.W.; Armstrong, A., Eds.; American Chemical Society: Washington, D.C., 1992; *ACS Symposium Series*, *484*, pp 429-462.
27. Ponnampalam, R.; Mondy, N. I. *J. Agric. Food Chem.* **1983**, *31*, 493-495.

28. Powell, R. D.; Brewer, T. A.; Dunn, J. W.; Carson, J. M.; Cole, R. H. *Potential for Storing Chipping Potatoes in Pennsylvania*; University of Pennsylvania Marketing Research Report 6, AE and RS 198, 1989.
29. Takagi, K.; Toyoda, M.; Fujiyama, Y.; Saito, Y. *J. Food Hygienic Soc. Jpn.* **1990**, *31*, 67-73.
30. Friedman, M.; Rayburn, J. R.; Bantle, J. A. *Food and Chem. Toxicol.* **1991**, *28*, 537-547.
31. Friedman, M.; Rayburn, J. R.; Bantle, J. A. *J. Agric. Food Chem.* **1992**, *40*, 1617-1624.
32. Stapleton, A.; Allen, P. V.; Friedman, M.; Belknap, W. R. *J. Agric. Food Chem.* **1991**, *39*, 1187-1203.
33. Stapleton, A.; Allen, P. V.; Tao, H. P.; Belknap, W. R.; Friedman, M. *Protein Expression Purif.* **1992**, *3*, 85-92.
34. Stapleton, A.; Beetham, J. K.; Pinot, F.; Garbarino, J. E.; Rockhold, D. R.; Friedman, M.; Hammock, B. D.; Belknap, W. R. *Plant J.* **1994**, *6*, 251-258.
35. Hellenäs, K.-E.; Branzell, C.; Johnsson, H.; Slanina, P. *J. Sci. Food Agric.* **1995**, *67*, 125-128.
36. Friedman, M.; Levin, C. E. *J. Agric. Food Chem.* **1992**, *40*, 2157-2163.
37. Stanker, L. H.; Kampos-Holtzapple, C.; Friedman, M. *J. Agric. Food Chem.* **1994**, *42*, 2360-2366.
38. Sinden, S. L.; Deahl, K. L.; Aulenbach, B. B. *J. Food Sci.* **1976**, *41*, 520-523.
39. Zitnak, A.; Filadelfi, M. A. *J. Can. Inst. Food Sci. Technol.* **1985**, *18*, 337-339.
40. Kaaber, L. *Norw. J. Agric. Sci.* **1993**, *7*, 221-229.
41. Zitnak, A.; Filadelfi-Keszi, M. A. *J. Food Biochemistry* **1988**, *12*, 183-190.
42. Friedman, M.; McDonald, G. M. *J. Agric. Food Chem.* **1995**, *43*, 1501-1506.
43. Friedman, M.; McDonald, G.; Haddon, W. F. *J. Agric. Food Chem.* **1993**, *41*, 1397-1406.
44. Zitnak, A. *Am. Potato J.* **1981**, *58*, 415-421.
45. Petermann, J. B.; Morris, S. C. *Plant Sci.* **1985**, *39*, 105-110.
46. Dale, M. F. B.; Griffiths, D. W.; Bain, H.; Todd, D. *Ann. Appl. Biol.* **1993**, *123*, 411-418.
47. Mondy, N. I.; Gosselin, B. *J. Food Sci.* **1988**, *53*, 756-759.
48. Mondy, N. I.; Leja, M.; Gosselin, B. *J. Food Sci.* **1987**, *52*, 631-635.
49. Olsson, K. *Potato Res.* **1986**, *29*, 1-12.
50. Fitzpatrick, T. J.; Herb, S. F.; Osman, S. F.; McDermott, J. A. *Am. Potato J.* **1977**, *54*, 539-544.
51. Tingey, W. M. *Am. Potato J.* **1984**, *61*, 157-164.
52. Olsson, K. *J. Phytopathol.* **1987**, *118*, 347-357.
53. Olsson, K. *Impact Damage, Gangrene and Dry Rot in Potato - Important Biochemical Factors in Screening for Resistance and Quality in Breeding Material.* Ph. D. Thesis, The Swedish University of Agricultural Science, Svalov, 1989.
54. Mondy, N. I.; Chandra, S. *HortSci.* **1979**, *14*, 173-174.
55. Burton, W. G. The Potato. Longman Scientific: Harlow, England, **1989**.

56. Stoddard, L. M. *Glycoalkaloid Synthesis During Storage of Potatoes.* B.S. Thesis, Procter Department of Food Science, The University of Leeds, UK, 1992.
57. Sowokinos, J. In *The Molecular and Cellular Biology of the Potato*; Vayda, M. J.; Park, W. D., Eds.; CAB International: Wallingford, UK, 1990; pp 137-158.
58. Van Berkel, J.; Salamini, F.; Gebhardt, C. *Plant Physiol.* **1994**, *104*, 445-452.
59. Corsini, D. L.; Pavek, J. J.; Dean, B. *Am. Potato J.* **1992**, *69*, 423-435.
60. Lisinska, G.; Leszczynski, W. In *Potato Science and Technology*; Elsevier Applied Science: London and New York, **1989**; pp 129-164.
61. Pavek, J. J.; Brown, C. R.; Martin, M. W.; Corsini, D. L. *Am. Potato J.* **1993**, *70*, 43-48.
62. Friedman, M. *J. Agric. Food Chem.* **1994**, *42*, 3-20.
63. Friedman, M.; Bautista, F. F. *J. Agric. Food Chem.* **1995**, *43*, 69-76.
64. Friedman, M.; Molnar-Perl, I.; Knighton, D. *Food Additives Contam.* **1992**, *9*, 499-503.
65. Molnar-Perl, I.; Friedman, M. *J. Agric. Food Chem.* **1990**, *38*, 1652-1656.
66. Ziderman, I. I.; Gregorski, K. S.; Lopez, S. V.; Friedman, M. *J. Agric. Food Chem.* **1989**, *37*, 1480-1486.
67. Sawant, D. W,; Dhumal, S. S.; Kadam, S. S. In *Potato: Production, Processing, Products*; Salunkhe, D. K.,; Kadam, S. S.; Jadhav, S. J., Eds.; CRC Press: Boca Raton, FL, 1991; pp 37-68.
68. Mondy, N. I.; Seetharaman, K. *J. Food Sci.* **1990**, *55*, 1740-1742.
69. Swallow, J.A. In *Nutritional and Toxicological Consequences of Food Processing*; Friedman, M., Ed.; Plenum: New York, 1991; pp 11-31.
70. Vaughn, S. F.; Spencer, G. F. *Am. Potato J.* **1991**, *68*, 821-831.
71. Kumar, A.; Jadhav, S. J.; Salunkhe, D. K. In *Potato: Production, Processing, Products*; Salunkhe, D. K.; Kadam, S. S.; Jadhav, S. J., Eds.; CRC Press: Boca Raton, FL, 1991; pp 247-267.
72. Brown, C. R. *Am. Potato J.* **1993**, *70*, 363-373.
73. Shetty, K. K.; Dwelle, R. B.; Fellman, J. K.; Patterson, M. E. *Potato Res.* **1991**, *34*, 253-260.

# SENSOR TECHNOLOGY
## AND ANALYTICAL METHODOLOGY

Chapter 18

# Periodate Oxidative Degradation of Amadori Compounds

## Formation of $N^\epsilon$-Carboxymethyllysine and N-Carboxymethylamino Acids as Markers of the Early Maillard Reaction

Raphaël Badoud, Laurent B. Fay, Fabienne Hunston, and Gudrun Pratz

Nestlé Research Centre, Nestec Limited, P.O. Box 44, Ver-chez-les-Blanc, CH—1000 Lausanne 26, Switzerland

Protein-bound and free Amadori compounds can be selectively degraded to their corresponding N-carboxymethylamino acids using periodic acid oxidation followed by acid hydrolysis. In the case of glycosylated lysine, $N^\epsilon$-carboxymethyllysine (CML) is formed. CML can easily be determined by HPLC after pre-column derivatization with o-phthalaldehyde (OPA) and 3-mercaptopropionic acid while other N-carboxymethylamino acids can be analyzed by GC or GC-MS after suitable derivatization. The amount of CML formed gives a direct measure of the blockage of protein lysino groups and thus provides a measure of the extent of early Maillard reaction. This approach can be applied both to food products and to biological materials.

The reaction of reducing sugars with amino acids or free amino groups in proteins is known as the Maillard reaction or non-enzymatic browning. This reaction covers a whole range of complex transformations which lead to the formation of numerous volatile and non-volatile compounds. The first consistent description of the chemical pathways of the Maillard reaction was proposed by Hodge in 1953 (*1*) and this scheme is still used today. The Maillard reaction can be divided into three major phases (*2*): (i) an early stage consisting of the formation of Amadori and Heyns compounds as a result of the condensation of primary amino groups of amino acids, peptides or proteins with the carbonyl group of reducing sugars; (ii) a second, advanced stage, leading to the formation of degradation products, reactive intermediates and volatile compounds (formation of flavor); and (iii) the final stage, characterized by the production of nitrogen-containing brown polymers and co-polymers known as the melanoidins.

As far as food products are concerned, the most important and obvious consequences of the Maillard reaction are the development of color and flavor. These effects may or may not be desirable, but they are always accompanied by a decrease in the nutritional value of proteins. In milk for instance, the reaction of lactose with ε-lysine residues results in the formation of the Amadori product ε-deoxylactulosyllysine and renders this essential amino acid biologically unavailable (*3*).

Non-enzymatic browning reactions also occur in biological systems where the physiological properties of certain long-lived proteins of the human body (hemoglobin, glycogen, lens crystallin, albumin, etc.) may be damaged during aging and particularly in diabetic patients (*4*). This is more often referred to as non-enzymatic glycosylation or glycation and mainly involves the ε-lysino groups of these proteins, although N-terminal amino groups may also be affected. Reliable methods to measure the extent and sites of glycosylation are therefore required.

**Methods Available for Measuring Non-Enzymatic Glycosylation**

Several methods have been developed for assaying non-enzymatic glycosylation. As far as biological systems are concerned, these have been extensively reviewed by A. J. Furth in 1988 (*5*). They include both assays on intact proteins after chemical degradation and selective detection of e.g. 5-hydroxymethylfurfural (HMF) and formaldehyde using the thiobarbituric assay (TBA), and assays on protein hydrolysates with or without previous reduction of the protein-bound Amadori compound. In this last case, the analysis is based on the determination of furosine which is specifically formed from lysine Amadori compounds with a yield of approximately 30% (*6*). The furosine method, originally developed for milk (*7*), has been the subject of several analytical improvements both for food products (*8*) and biological materials (*9*). More recently, another method has been proposed to evaluate the extent of early Maillard reaction in milk products. This method is based on direct measurement of the Amadori product lactuloselysine which is released after complete enzymatic hydrolysis (*10*).

**Materials and Methods**

The Amadori compounds $N^\alpha$-acetyl-$N^\varepsilon$-(1-deoxyfructos-1-yl)-L-Lysine and $N^\alpha$-formyl-$N^\varepsilon$-(1-deoxylactulos-1-yl)-L-lysine, and $N^\varepsilon$-carboxymethyllysine (CML) were prepared as previously described (*11, 12*). Glucose- and galactose-glycosylated lysozyme were prepared as follows (*13*) : 2 g of lysozyme (from hen's egg white, Fluka Nr. 62971) and 600 mg of either glucose or galactose were dissolved in 10 mL of distilled water and the pH was adjusted to 8.5 with 5% triethylamine. The solution was then freeze-dried and the resulting powder equilibrated to a water activity of 0.33 using saturated magnesium chloride for several days at room temperature. The samples were then stored at 47°C for 2, 24, 48 and 120 hours to induce maillardization. The glycosylated samples (120 mg) were treated overnight with 16 mL of 15 mM aqueous periodic acid. Excess reagent was destroyed by the addition of 400 µL of 2N sodium thiosulfate. Water (45 mL) and concentrated hydrochloric acid (60 mL) were added, and the mixture heated at 110°C for 24 hours. Aliquots were evaporated to dryness and reconstituted in water prior to HPLC analysis.

Food products (milk and tomato powders, pasta) were treated as described above. Typically, 150 mg of finely divided sample were mixed with 50 mL of 50 mM

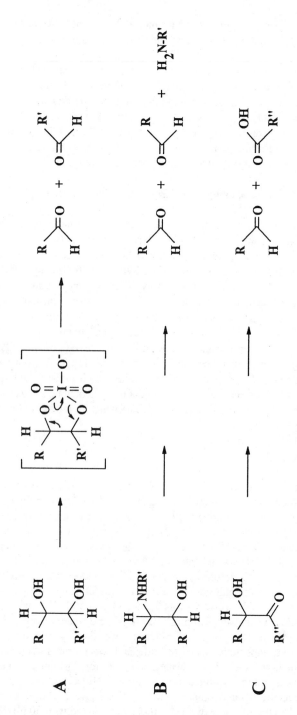

Figure 1. Periodate oxidation of 1,2-diols (A), 1,2-amino alcohols (B) and α-hydroxy ketones (C).

periodic acid and 0.5 mL of sodium dodecylsulfate (to improve wettability). The mixture was well shaken and stored overnight in the dark. Two mL of 2N sodium thiosulfate were added followed by 25 mL of water and 75 mL of 12N HCl. The mixture was then hydrolyzed as usual at 110°C for 24 hrs.

**HPLC Analysis.** CML and lysine were determined by HPLC after pre-column derivatization with o-phthalaldehyde (OPA) and 3-mercaptopropionic acid using the AminoQuant software (Hewlett-Packard) and a model HP 1090M (Hewlett-Packard) chromatograph equipped with an automated sample processor and injector, a column oven, a derivatization micro-oven and a model HP 1046A fluorescence detector. The OPA reagent was prepared from 25 mg OPA (Serva or Fluka), 500 μL methanol, 3.25 mL water, 1.25 mL 1M borate buffer (pH 10.4, Pierce) and 25 μL 3-mercaptopropionic (Fluka). The separation was performed through a 125 x 4 mm I.D. column (Stagroma) filled with Hypersil ODS 3 μm (Shandon). The mobile phase was composed of A) 25 mM sodium acetate buffer pH 7.2 containing 0.7% tetrahydrofuran and B) 100 mM sodium acetate buffer pH 7.2 in acetonitrile (1:4 v/v). The gradient was 0-30% B/0-9 min; 30-50% B/9-11 min; 50-100% B/13-14 min; 100% B/14-18 min; 100-0% B/18-19 min. The flow rate was 0.8 mL/min and the column was thermostated at 40°C. The fluorescence detector was set at 230 nm (excitation) and 450 nm (emission).

**Periodate Oxidative Cleavage of 1,2-Diols and Free Amadori Compounds**

The carbon-carbon bond of 1,2-diols, 1,2-amino alcohols, α-hydroxy ketones and related compounds can easily be cleaved using periodate ions to generate carbonyl compounds as shown in Figure 1. The Amadori compound formed by rearrangement of the intermediate glysosylamine is present in solution mainly in the β-pyranosyl form (Figure 2). This cyclic form allows the 1,2-diol (formally an α-hydroxy hemiacetal) on the C(2)-C(3) bond to have the most favorable *cis*-configuration required for periodate oxidation. The Amadori compound also bears a secondary α-amino alcohol function on the C(1)-C(2) bond which can be cleaved under neutral or basic conditions. However, when treating the Amadori compounds $N^\alpha$-acetyl-$N^\epsilon$-(1-deoxyfructos-1-yl)-L-lysine (A-DFL) (Figure 3, R=H, R'=acetyl) or $N^\alpha$-formyl-$N^\epsilon$-(1-deoxylactulos-1-yl)-L-lysine (F-DLL, R=galactosyl, R'=formyl) with periodic acid at pH < 4, it was observed that the reaction could be selectively directed towards cleavage of the C(2)-C(3) bond. This gives rise to the formation, presumably via an intermediate labile α-aminoester, of $N^\epsilon$-carboxymethyllysine (CML) which is obtained in a yield of approximately 90%. In these acidic conditions, it is likely that protonation of the secondary amino group blocks formation of the cyclic periodate ester intermediate and thus hinders cleavage of the C(1)-C(2) bond. CML is also formed in food products (*14*) and biological materials as a result of *in situ* oxidation of lysine Amadori products or, as proposed recently (*15*), by the reaction of glyoxal (an autoxidation product of glucose formed under physiological conditions) with lysine.

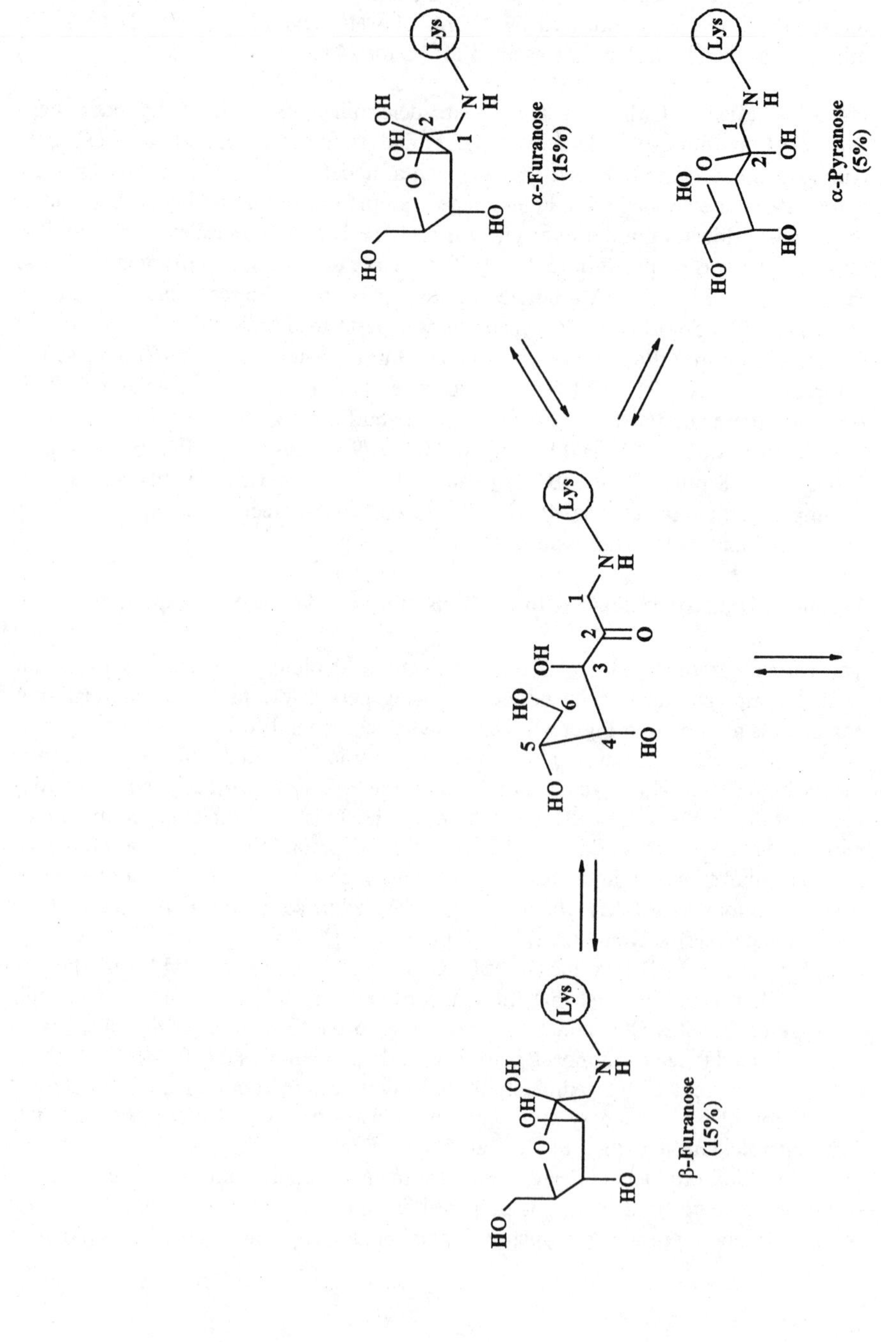

Lys

N—H

α-Furanose
(15%)

Lys

N—H

α-Pyranose
(5%)

Lys

N—H

Lys

N—H

β-Furanose
(15%)

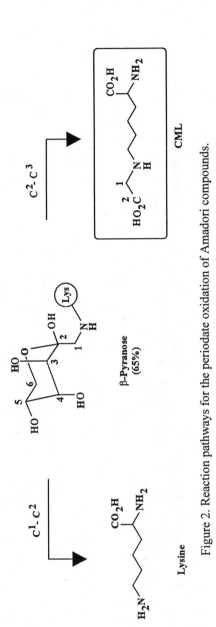

Figure 2. Reaction pathways for the periodate oxidation of Amadori compounds.

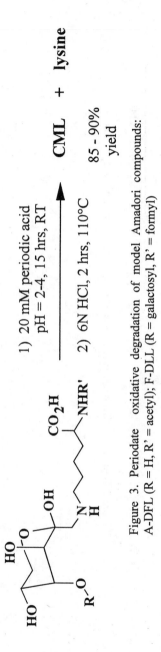

Figure 3. Periodate oxidative degradation of model Amadori compounds: A-DFL (R = H, R' = acetyl); F-DLL (R = galactosyl, R' = formyl)

## Periodate Oxidation of Protein-Bound Amadori Compounds: Development of the CML Method

The approach, which consists of the oxidative cleavage of free Amadori compounds, was further applied to protein-bound Amadori compounds and to more complex systems. The results of periodate oxidation of lysozyme, chosen as a model protein, and previously incubated for various lengths of time with both glucose or galactose, are presented in Figure 4. CML and lysine were analyzed after acid hydrolysis. The sum of CML and lysine is approximately constant and the ratio CML/(CML + lysine) can be directly used to estimate the extent of blocked lysine. Thus, it is assumed that conversion of the glycosylated protein-bound lysine residues to CML is practically quantitative. Evidence for this is illustrated in Figure 5. This shows the CML and lysine content of a milk powder to which increasing amounts of pure A-DFL were added. The CML content increases linearly as a function of the amount of free Amadori compound added, whereas the lysine concentration remains almost constant. Therefore, the periodate oxidation of A-DFL, performed under these specific conditions, results in the formation of CML exclusively. This result confirms that, as mentioned above, the amount of CML and lysine found in the hydrolysate can be directly used to calculate the extent of lysine blockage.

In a previous paper (*11*), it was shown that estimation of blocked lysine in milk powders using the CML method is in excellent agreement with the furosine method. This information is essential for assessing the nutritional quality of food proteins. When used in this context, and not only as a tool to monitor heat treatment or aging, the furosine method needs correction factors to account for its yield during acid hydrolysis (*6*). In comparison, the CML approach is more straightforward. Furthermore, the presence of CML in a sample hydrolysate can convey different information. As recently discussed by Ruttkat *et al.* (*14*) CML can be considered as a better indicator of oxidative glycosylation in both food and biological samples in comparison to furosine as it is more stable. This marker is even more reliable if the formation of additional CML as an artifact during acid hydrolysis is prevented using preliminary reduction with sodium borohydride (*16*). In the present work, the CML which is formed as a result of periodate oxidation reflects the extent of lysine glycosylation. It is not only an indicator of heat damage. Like the furosine method, it allows to measure the proportion of lysine groups nutritionally unavailable. The advantage of the CML method is that a single and very sensitive analytical method is used to assess both oxidative glycosylation and the formation of protein-bound Amadori compounds. Finally, CML can be considered as a specific marker of the early stage of the Maillard reaction. It provides a valuable tool for early detection of the reaction of reducing sugars with ε-lysino groups. Applications of this method to monitor chemical changes in various food products during processing and storage are presented thereafter.

### Applications to Various Food Products

Milk powders call for attention because of their importance in manufacturing products with a high nutritional value. However, many other food products,

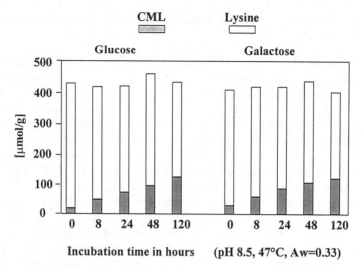

Figure 4. Glycosylation of lysozyme in the presence of glucose and galactose for various lengths of time. Determination of CML and lysine in the acid hydrolysate after periodate oxidation.

particularly dehydrated vegetables and pasta, can be assayed using the CML method. This can be applied in accelerated storage tests to predict the shelf-life of raw materials and end-products. As an example, Figure 6 shows the HPLC amino acid profile of a dry pasta hydrolysate, with (top chromatogram) and without (bottom chromatogram) previous oxidative treatment with periodic acid. The presence of CML eluting between serine (Ser) and histidine (His) is clearly visible in the treated sample. Figure 7 shows the result of the determination of CML and lysine in various tomato powders stored between 0 and 10 weeks under specific test conditions. The sum of CML and lysine (not explicitly shown) tends to decrease with time due to partial destruction of lysine under advanced Maillard conditions.

**Applications to other Glycosylated Amino Acids**

Application of the periodate oxidation approach to the determination of other glycosylated amino acids such as free and N-terminal amino acids, has been demonstrated both for food products (*12*) and in human hemoglobin (*17*). In these cases, specific N-carboxymethylamino acids are analyzed by GC or GC-MS after suitable derivatization using ethyl chloroformate or as their O-isobutyl N-pentafluoropropionyl derivatives. For instance, N-carboxymethylvaline was detected in a hemoglobin hydrolysate which confirms that valine is the N-terminal amino acid and that it exists partly in a glycosylated form. Furthermore, the level of N-carboxymethylvaline in a sample from a diabetic patient was found to be twice as high when compared to a sample from a normal patient. Similarly, the level of CML in the samples from diabetic *vs.* normal patient was in a ratio of 4:1.

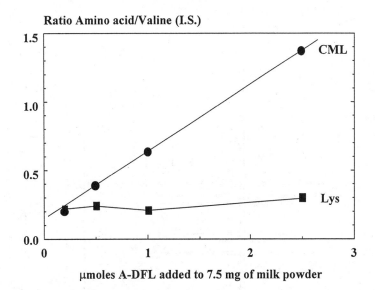

Figure 5. CML and lysine content in a model milk powder hydrolysate. Increasing quantities of pure A-DFL were added to the starting material prior to oxidation with periodic acid. Values correspond to the ratio CML (●) or lysine (■) to valine. Valine was chosen as an intrinsic internal standard.

**Conclusion**

Periodate oxidation of protein-bound Amadori compounds allows an estimation of the extent of the early Maillard reaction in various food and biological samples. The main features of this method are the following: (i) the same oxidation product is formed regardless of the nature of the reducing sugar (glucose, lactose or galactose); (ii) carbohydrates, either free or bound through glycosidic linkages do not interfere; (iii) glycosylation of several free amino acids and N-terminal amino acids in peptides and proteins can be assayed; (iv) sample preparation for CML is straightforward: the crude sample is sequentially oxidized then hydrolyzed in the same vessel and assayed by HPLC after automated OPA derivatization; (v) other N-carboxymethylamino acids can be analyzed by GC or GC-MS after suitable derivatization; (vi) the method can be applied to complex biological materials as a tool to identify the sites and the extent of protein glycosylation.

**Acknowledgments.**  The authors are grateful to Mrs. E. Prior for critically reading the manuscript.

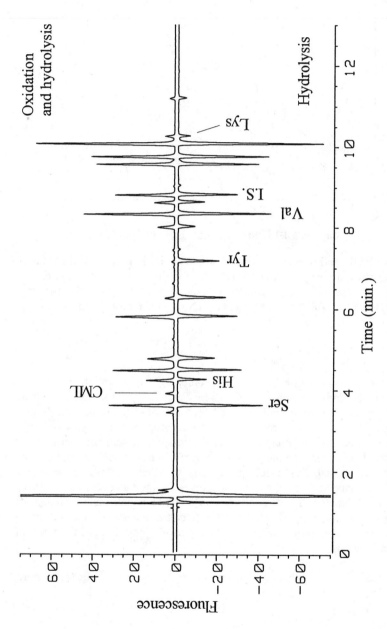

Figure 6. HPLC chromatogram of the acid hydrolysate of a dry pasta product which was previously treated or not with periodic acid. The position of various other protein amino acids are shown. I.S.: internal standard = norvaline.

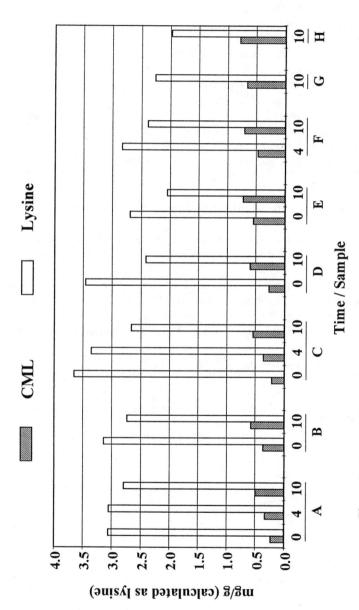

Figure 7. CML and lysine content of tomato powders of various origins (A to H), stored for 0, 4 and 10 weeks under controlled test conditions.

220     CHEMICAL MARKERS FOR PROCESSED AND STORED FOODS

## Literature cited

1. Hodge, J. E. *J. Agric.Food Chem.* **1953**, *1*, 928-943
2. Mauron, J. *Prog. Fd. Nutr. Sci.* **1981**, *5*, 5-35
3. Finot, P. A., Deutsch, R. and Bujard, E. *Prog. Fd. Nutr. Sci.* **1981**, *5*, 345-355
4. Cerami, A. In *Maillard Reactions in Chemistry, Food, and Health;* Labuza, T.P., Reineccius, G.A., Monnier, V., O'Brien, J. and Baynes, J.W., Ed.; The Royal Society of Chemistry: London, 1994, pp 1-10
5. Furth. A. J. *Anal. Biochem.* **1988**, *175*, 347-360
6. Finot, P. A. and Mauron, J. *Helv. Chim. Acta* **1969**, *52*, 1488-1495; Schleicher, E. and Wieland, O. H., *J. Clin.Chem. Clin. Biochem.*, **1981**, *19*, 81-87
7. Bujard, E. and Finot, P. A. *Ann. Nutr. Alim.* **1978**, *32*, 291-305
8. Resmini, P., Pellegrino, L. and Batelli, G. *Ital. J. Food Sci.* **1990**, *3*, 173-183
9. Wu, Y. C., Monnier, V. and Friedlander, M. *J. Chromatogr. B*, **1995**, *667*, 328-332
10. Henle, T., Walker, H. and Klostermeyer, H. *Z. Lebensm. Unters. Forsch.* **1991**, *193*, 119-122
11. Badoud, R., Hunston, F., Fay, L. and Pratz, G. In *The Maillard Reaction in Food Processing, Human Nutrition and Physiology*; Finot, P. A., Aeschbacher, H. U, Hurrel, R. F. and Liardon, R., Ed.; Birkhäuser: Basle, 1990, pp 79-84
12. Badoud, R., Fay, L., Richli, U. and Husek, P. *J. Chromatogr.* **1991**, *552*, 345-351
13. Wu, S., Govindarajan, S., Smith, T., Rosen , J.D. and Ho C.T. In *The Maillard Reaction in Food Processing, Human Nutrition and Physiology;* Finot, P.A., Aeschbacher H.U., Hurrel, R.F. and Liardon, R., Ed.; Birkhäuser: Basle, 1990, pp 85-90
14. Ruttkat, A. and Erbersdobler, H. F. *J. Sci. Food Agric.* **1995**, *68*, 261-263
15. Wells Knecht, K. J., Zyzak, D. V., Litchfield, J. E, Thorpe, S. R. and Baynes, J. W. *Biochemistry*, **1995**, *34*, 3702-3709
16. Hartkopf, J., Pahlke, C., Lüdemann G. and Erbersdobler, H. F. *J. Chromatogr.* **1991**, *552*, 345-351
17. Badoud, R. and Fay, L. *Amino Acids*, **1993**, *5*, 367-375

# Chapter 19

# Haptenic Sugar Antigens as Immunochemical Markers for the Maillard Reaction of Processed and Stored Milk Products

T. Matsuda[1] and Y. Kato[2]

[1]Department of Applied Biological Sciences, School of Agricultural Sciences, Nagoya University, Chikusa-ku, Nagoya 464-01, Japan
[2]Department of Clinical Nutrition, Kawasaki University of Medical Welfare, 288 Matsushima, Kurashiki, Okayama 701-01, Japan

Milk proteins are chemically modified by amino-carbonyl reaction (Maillard reaction) with lactose during processing and storage. We have obtained rabbit antisera and mouse monoclonal antibodies against the Maillard reaction products between lactose and protein amino groups. These antibodies specifically recognize the lactose-lysine reaction products as haptenic sugar antigens, and are useful to detect the Maillard reaction products in processed and stored milk samples. Some market milk samples were boiled for 30 min, while several powdered milk products were kept at 25 °C for a few weeks. The Maillard reaction products in these samples were analyzed by an enzyme immunoassay ELISA. The results indicated that the Maillard products already existed in these market milk and milk products and were further produced by heating of the market milk samples and storage of the powdered milk samples.

Proteins are chemically modified with reducing sugars by Maillard reaction between amino- and carbonyl-groups during processing, storage and cooking of various food (1). The Maillard reaction between sugars and proteins as food constituents affect the food quality, e.g., reduction of protein nutritional value, desirable or undesirable browning of food, alteration of functional properties of food proteins.

The Maillard reaction of lactose and proteins in milk has long been investigated mainly from a view point of nutritional quality of pasteurized and powdered milk or milk products (2). Availability of an essential amino acid, lysine, is decreased by the reaction with lactose during heating for pasteurization and spray-drying for powdering of milk. Loss of available lysine and production of reaction intermediates are estimated by amino acid analysis of acid hydrolysates of milk samples.

Immunochemical methods using specific antibodies are now widely used for analyses to detect a specific component in complex mixtures of food components. The authors have found that novel antigenic structures are produced by Maillard reaction of proteins with lactose (3) and obtained rabbit polyclonal and mouse monoclonal antibodies which specifically recognize lactose-protein Maillard reaction products (4,5). These antibodies reacted not only with lactose-milk protein reaction products but also with lactose-egg protein reaction products, indicating that these antibodies mainly recognize lactose-derived antigenic components in the Maillard

0097–6156/96/0631–0221$15.00/0

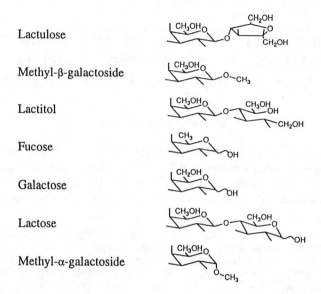

Lactulose

Methyl-β-galactoside

Lactitol

Fucose

Galactose

Lactose

Methyl-α-galactoside

Figure 1. Sugars and sugar derivatives used as competitors for the competitive inhibition ELISA.

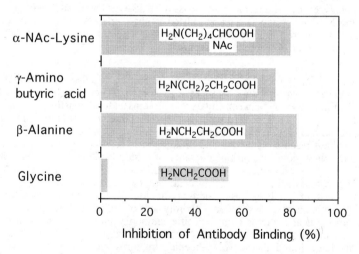

Figure 2. Inhibition of the antibody binding by Maillard reaction products of lactose with some amino acids.    The lactose-amino acid Maillard products were used as competitors against binding of the monoclonal antibody L101 to lactose-protein Maillard products.

reaction products.   Such lactose-derived antigens could be an immunochemical marker to evaluate quality of market milk and powdered milk products concerning Maillard reaction caused by processing and/or storage (*6*).   In the present study we investigated the antigen binding specificity of one monoclonal antibody, L101, specific for the lactose-protein Maillard products, and applied the monoclonal antibody to estimate Maillard products in various milk samples.

## Epitope structure recognized by the specific antibody

The products of Maillard reaction between lactose and protein amino groups are complex mixtures of many unidentified compounds.   Among them, a relatively stable and dominant intermediate compound is ε-*N*-deoxylactulosyl-lysine which is the Amadori rearrangement product (*7*).   To estimate an epitope structure recognized by the monoclonal antibody L101, competitive inhibitory activity of several sugars (Figure 1) against the antibody binding to the Maillard product were examined by competitive ELISA (*5,8*).   The strongest inhibitor was lactulose, which is a structural analog of the Amadori rearrangement product of the lactose-amino group Maillard reaction.   Methyl-β-galactoside, lactitol and lactose showed inhibitory activity but their effectiveness was much lower than that of lactulose.   Methyl-α-galactoside, galactose and fucose showed no inhibitory activity under the experimental condition used.   Lactulose has been reported to be produced from lactose in large amount during the heating of milk (*9*), and such lactulose would interfere the antibody binding to the haptenic sugar antigens in milk.   However, free lactulose in heated milk samples can be eliminated from the solid-phase immunoassay system such as ELISA, because only protein-bound sugar antigens but not free lactulose can be adsorbed on the surface of the ELISA plate.   Therefore, the lactose-protein Maillard products can be analyzed immunochemically without interference of lactulose produced in heated milk products.

   Contribution of amino acid moiety to the epitope structure was estimated by the competitive ELISA using the Maillard reaction products of lactose with α-*N*-acetyl-*L*-lysine, γ-aminobutylic acid, β-alanine and glycine.   As shown in Figure 2, the Maillard reaction products of lactose with α-*N*-acetyl-lysine, γ-aminobutylic acid, β-alanine strongly inhibited the antibody binding, whereas the product of glycine with lactose showed no inhibitory activity.   This indicates that not only amino group but also a part of alkyl chain of the lysine residue contribute the epitope structure of the lactose-protein Maillard product.   The monoclonal antibody, L101, is also suggested to be raised against the Maillard product of lactose with ε-amino group of lysine residues but not against the product with α-amino group of the *N*-terminal amino acid residue of proteins.

## Lactose-derived sugar antigens in commercial milk products

Maillard reaction of lactose with milk proteins proceeds during processing such as pasteurization and drying of milk.   Apparent content of lactose-derived haptenic sugar antigens in several market milk and powdered skim milk samples was estimated by ELISA using the monoclonal antibody L101.   Appropriately diluted or dissolved milk samples were directly used as antigen solution for coating of ELISA plates.   Figure 3 shows ELISA values of several representative milk samples.   The powdered milk samples showed relatively higher ELISA value, indicating their higher content of the lactose-derived haptenic antigen.   Among pasteurized market milk samples tested, UHT-milk pasteurized at 140 °C for 3 sec showed stronger positive reaction, and UHT-milk pasteurized at 140 °C for 2 sec and HTST milk pasteurized at 85 °C for 15

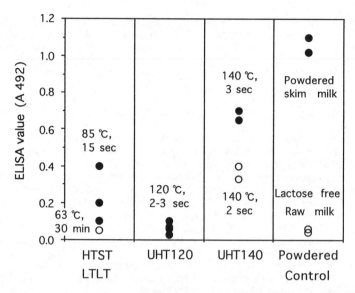

Figure 3. Immunochemical detection of lactose-protein Maillard products in several market and powdered milk samples. ELISA plates were coated with appropriately diluted milk samples and the lactose-derived antigens were detected with L101. Apparent antigen amount is shown as ELISA value (absorbance at 405 nm).

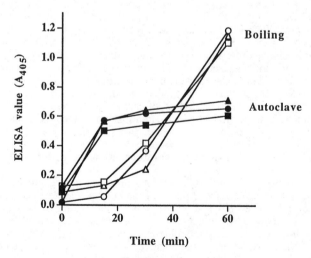

Figure 4. Increase in Maillard reaction products by heating in market milk samples. Three kinds of market milk samples were heated in boiling water (open symbol) or autoclaved at 120 °C (closed symbol) for 60 min. Apparent antigen amount estimated with L101 is shown as ELISA value (absorbance at 405 nm).

sec showed moderate positive reaction.   On the other hand, LTLT-milk (63 °C, 30 min) and UHT-milk pasteurized at 120 °C showed very weak or almost no ELISA value.   Raw milk and a lactose-free milk sample were also negative to the antibody binding.   Both temperature and time for pasteurization appears to affect the content of lactose-derived sugar antigens in market milk samples, and concentration and spray-drying processes might accelerate the Maillard reaction between lactose and milk proteins leading to the production of the specific haptenic antigens.

**Production of the sugar antigens by heating and storage**

To determine whether Maillard reaction between lactose and proteins proceeds during storage and home-cooking, several commercial milk samples were boiled, autoclaved or stored and, then, production of the lactose-derived sugar antigens was monitored by ELISA.   Three kinds of UHT-pasteurized milk samples were heated in boiling water for 10 to 60 min.   The apparent antigen content in each milk sample gradually increased with increase in heating time, especially after heating for 30 min (Figure 4).   The same market milk samples were autoclaved at 120 °C for 60 min.   The ELISA value also increased by the autoclave heating, but no large increase was observed after the autoclave for 20 min (Figure 4).   The ELISA value of the market milk samples boiled for 60 min was about twice that of the same samples autoclaved for 20 min, and much higher than those of the samples without boiling or autoclaving.   There were no large difference in the ELISA value among the milk samples pasteurized at 120, 130 and 140 °C.   These results suggest that the Maillard reaction of milk proceeds not only by industrial pasteurization, concentration and drying but also by home-cooking.

Some powdered milk samples were kept at 25 °C for 4 weeks, and production of the specific antigen was also monitored by ELISA.   An typical result of a skim milk sample stored for 4 weeks is shown in Figure 5.   The ELISA value gradually increased with the storage period.   However, the increase in ELISA value induced by

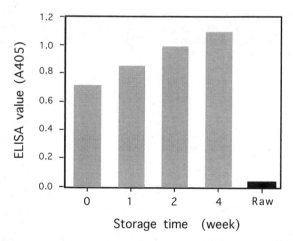

Figure 5. Increase in Maillard reaction products by storage in a powdered skim milk sample.   A powdered skim milk sample was kept at 25 °C for 4 weeks. Apparent antigen amount estimated with L101 is shown as ELISA value (absorbance at 405 nm).

the storage is considerably lower than that induced by industrial processing.    These results agree with those obtained previously by other analytical methods.    Thus, the lactose-derived sugar antigens would be useful immunochemical markers for the quality of processed and stored milk products.

## Literature Cited

(1)    Reynolds, T.H. Cemistry of nonenzymatic browning. I. The reaction between aldoses and amines. *Adv. Food Res.* 1963, *12*, 1-52.

(2)    Finot, P. A., Deutsch, R. and Bujard, E. The extent of Maillard reaction during the processing of milk. In *Maillard reaction of food*, Eriksson, C., Ed., Pergamon Press, Oxford, 1981 pp 345-355.

(3)    Matsuda, T., Nakashima, I., Kato, Y. and Nakamura, R. Antibody response to haptenic sugar antigen: immunodominancy of protein-bound lactose formed by amino-carbonyl reaction. *Mol. Immunol.* 1987, *24*, 421-425.

(4)    Matsuda, T., Kato, Y., Watanabe, K. and Nakamura, R. Direct evaluation of β-lactoglobulin lactosylation in early Maillard reaction using an antibody specific to protein-bound lactose. *J. Agric. Food Chem.* 1985, *33*, 1193-1196.

(5)    Matsuda, T., Ishiguro, H., Ohkubo, I., Sasaki, M. and Nakamura, R. Carbohydrate binding specificity of monoclonal antibodies raised against lactose-protein Maillard adducts. *J. Biochem.* 1992, *111*, 383-387.

(6)    Kato, Y., Matsuda, T., Kato, N. and Nakamura, R. Analysis of lactose-protein Maillard complexes in commercial milk products by using the specific monoclonal antibody. In *Maillard reaction in Chemistry, Food and Health* . Labuza, T. P. and Reineccius, G. A. Eds., Birkhauser Verlag, Basel, Boston, Berlin, 1995, pp188-194.

(7)    Ledl, F. Chemical pathways of the Maillard reaction. In *The Maillard Reaction in Food Processing, Human Nutrition and Physiology.*    Finot, P.A., Aeschbacher, H.U., Hurrell, R.F., Liardon, R. Eds. Birkhauser Verlag, Basel, 1990, pp 19-42.

(8)    Engvall, E. and Perlmann, P. Enzyme-linked immunosorbent assay (ELISA): Quantitative assay of immunoglobulin G. *Immunochemistry* 1971, *8*, 871-874.

(9)    Adachi, S. Formation of lactulose and tagatose from lactose in strongly heated milk. *Nature* 1958 *181*, 840-841.

Chapter 20

# Inactivation of Egg Trypsin Inhibitors by the Maillard Reaction

## A Biochemical Marker of Lysine and Arginine Modification

Y. Kato[1] and T. Matsuda[2]

[1]Department of Clinical Nutrition, Kawasaki University of Medical Welfare, 288 Matsushima, Kurashiki, Okayama 701−01, Japan
[2]Department of Applied Biological Sciences, School of Agricultural Sciences, Nagoya University, Chikusa-ku, Nagoya 464−01, Japan

Ovomuocids of chicken and Japanese quail eggs inhibit trypsin by forming a stable enzyme-inhibitor complex through the arginine and lysine residues at their reactive sites, respectively. Chemically modified arginine and lysine residues of the two ovomucoids by the Maillard reaction with reducing sugars were determined by the Sakaguchi's method and the fluorometric method using fluorescamine, respectively. The decrease in trypsin inhibitory activities of chicken ovomucoid was faster than that of quail ovomucoid, and the activity loss of chicken and quail ovomuocids were corresponding to the decrease in their free guanidino and amino groups, respectively. No activity loss of chicken ovomucoid was induced by the reaction with maltose, whereas quail ovomukoid was markedly inactivated, suggesting that the arginine residue is not modified directly with the carbonyl group of reducing sugars but with some active compounds degraded from the reducing sugars at the later stages of the Maillard reaction.

Food proteins are often denatured and/or chemically modified during heating, drying and storage of food. One of common protein modification events is Maillard reaction (*1, 2*), in which protein side chains react with carbonyl groups of reducing sugars. The initial step of this reaction is that the primary ε-amino groups of lysine side chains preferentially react as nucleophiles with carbonyl groups, resulting in the formation of stable glycosylamine intermediates such as Amadori compounds, which are detected in various processed and stored foods (*3*). The Amadori compounds are degraded to the deoxyosones, reactive α-dicarbonyl compounds, in neutral and acidic pH (*3*). Such reactive compounds interact with various food components including proteins, leading to formation of complex Maillard products.

Modification of food proteins with reducing sugars through Maillard reaction has been evaluated by various analytical methods; determination of free amino groups, amino acid analysis after acid hydrolysis, detection of brownish pigments and fluorescent compounds. Some Maillard products have also been analyzed immunochemically using specific antibodies.

Egg white is composed of about 10% protein and 0.5% glucose. Maillard reaction of egg white proteins with glucose proceeds during the production of dried egg white powder, resulting in the formation of brownish pigments and the alteration of powdered protein functional properties such as solubility, emulsifying activity , gelling property (*4*).

0097−6156/96/0631−0227$15.00/0

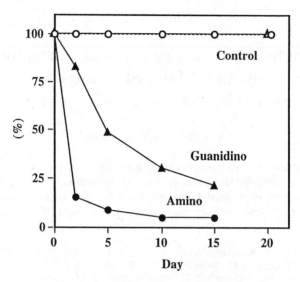

**Figure 1.** Decrease in free amino- (O) and guanidino- (Δ) groups of chicken ovomucoid by the incubation with glucose. The powdered ovomucoid was incubated with (closed symbol) or without (open symbol) glucose at 50 °C for 15 days.

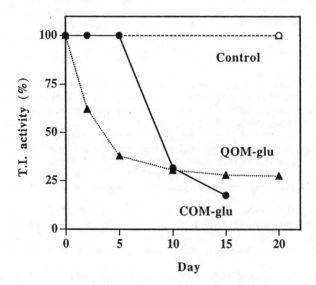

**Figure 2.** Decrease in trypsin inhibitory activity of chicken- (O) and Japanese quail- (Δ) ovomucoids by the incubation with glucose. The powdered ovomucoids were incubated with (closed symbol) or without (open symbol) glucose at 50 °C for 15 days.

About 10% of egg white proteins is a glycoprotein, ovomucoid(5). This glycoprotein with Mr of about 28,000 has been known to inhibit trypsin by forming stable enzyme-inhibitor complexes (6). The amino acid sequence around the inhibitory reactive site is well conserved among ovomucoids from various avian species, and amino acid residues of reactive sites are lysine or arginine (6). Chicken ovomucoid has one reactive site of arginine for trypsin, while Japanease quail ovomucoid has two reactive sites of lysine for trypsin. Chemical modification such as acetylation of lysine residues of Japanese quail ovomucoid leads to inactivation of its inhibitory activity against trypsin (7). Therefore, it seems likely that modification with reducing sugars through Maillard reaction also inactivate Japanese quail ovomucoid with lysine-type reactive sites. In the advanced stages of Maillard reaction, protein arginine residues are also modified probably with various degradation compounds derived from sugar-lysine reaction products (8). In the present study, we investigated decrease in trypsin inhibitory activity of ovomucoids by the Maillard reaction with reducing sugars and examined relationship between the biochemical activity and the other chemical markers of the Maillard reaction.

## Reaction of quail and chicken ovomucoids with glucose

**Free amino- and guanidino-groups.** As a model system of protein-sugar Maillard reaction, a powdered sample containing ovomucoid and glucose was incubated at 50 °C and 65% relative humidity for various periods from 2 to 20 days. The incubated samples were dissolved in 50 mM Tris-HCl buffer, pH 8.0 and used for chemical and biochemical analyses.

Free amino- and guanidino groups of incubated ovomucoids were determined by a fluorometric method using fluorescamine (9) and the method of Sakaguchi (10,11), respectively. A representative result on decrease in free amino- and guanidino-groups of chicken ovomucoid is shown in Figure 1. Free amino group rapidly decreased to 15% after the 2 day-incubation, and the residual free amino group was only a few percent after the 20 day-incubation. Free guanidino group also decreased but the reaction proceeded more slowly than that of amino group. About 50 and 25% of free guanidino group remained even after 5- and 15-day incubations, respectively. Similar results were obtained for Japanese quail ovomucoid. No decrease of amino and guanidino groups were observed for ovomucoids incubated in the absence of glucose.

**Trypsin inhibitory activity.** Trypsin inhibitory activity of ovomucoids were measured with α-N-benzoyl-L-arginine p-nitroanilide as a trypsin substrate (12) for samples incubated with glucose for various periods. As shown in Figure 2, the activity of Japanese quail ovomucoid decreased rapidly for the first five days of incubation, but about 25% of the original activity remained even after the 20 day-incubation. The decreasing profile of Japanese quail ovomucoid was relatively in good agreement with that of free amino group. On the other hand, chicken ovomucoid did not loose its trypsin inhibitory activity for the 5 day-incubation, even though free guanidino group decreased to about 50% after the 5 day-incubation. Then the activity rapidly decreased to about 30% after the 10 day-incubation.

It seems reasonable that TI activity loss of chicken ovomucoid did not correlate with the decrease in free amino group, because chicken ovomucoid has a reactive site of arginine. However, the TI activity loss did not necessarily correlate to the loss of the free guanidino group, especially for the first 5 days of incubation. This suggests that chemical modification of the reactive site arginine occurred with frequency lower than the other arginine residues of an ovomucoid molecule. Furthermore, Japanese quail ovomucoid which lost most of its free amino group still retained some extent of TI activity. Therefore, lysine or arginine at the reactive site of ovomucoid for trypsin inhibition might not easily be attacked by reducing sugars as compared with the these residues of the other part of a molecule, or ovomucoids modified with glucose might still retain weak affinity to trypsin.

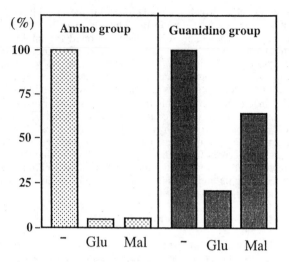

**Figure 3.**   Decrease in free amino- and guanidino-groups of chicken ovomucoid by the incubation with glucose (Glu) and maltose (Mal).    The powdered ovomucoid was incubated with (Glu, Mal) or without (-)sugars at 50 °C for 15 days.

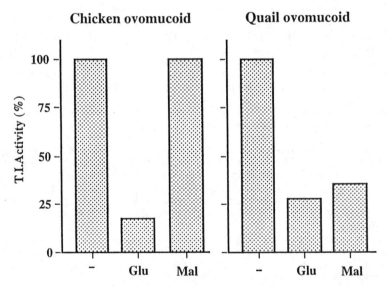

**Figure 4**.   Decrease in trypsin inhibitory activity of chicken- and Japanese quail-ovomucoids by the incubation with glucose (Glu) and maltose (Mal).    The powdered ovomucoids were incubated with (Glu, Mal) or without (-)sugars at 50 °C for 15 days.

## Reaction of ovomucoid with glucose and maltose

Since the structure of reducing sugars was reported to affect the progress of Maillard reaction (*13*), the reaction of ovmucoids with glucose and maltose was investigated and the effect of reaction on TI activity of chicken and Japanese quail ovomucoids was compared. The free amino group was rapidly decreased by the reaction not only with glucose but also with maltose (Figure 3).    The TI activity of Japanese quail ovomucoid was markedly decreased by the reaction with maltose, as well as glucose, during the incubation for 15 days (Figure 4).   On the other hand, the TI activity of chicken ovomucoid was not affected by the reaction with maltose, though the activity was decreased considerably by the reaction with glucose.    The rapid decrease in TI activity of Japanese quail ovomucoid by the reaction with maltose indicates that there is no large difference between glucose and maltose in the initial attacking efficiency against the reactive site lysine.    On the contrary, no effect of maltose on TI activity of chicken ovomucoid suggests that degradation of maltose-lysine reaction products to active compounds is much slower than that of glucose-lysine reaction products.

## Literature Cited

(1)  Reynolds, T.H. Cemistry of nonenzymatic browning. I. The reaction between aldoses and amines. *Adv. Food Res.* 1963, *12*, 1-52.

(2)  Namiki, M. Chemistry of Maillard reactions: Recent studies on the browning reaction mechanism and the development of antioxidants and mutagens. *Adv. Food Res.* 1988, *32*, 115-184.

(3)  Ledl, F. Chemical pathways of the Maillard reaction. in *"The Maillard Reaction in Food Processing, Human Nutrition and Physiology"* Finot, P.A., Aeschbacher, H.U., Hurrell, R.F., Liardon, R. Eds. Birkhauser Verlag, Basel, 1990, pp 19-42.

(4)  Kato, Y., Matsuda, T., Kato, N., Watanabe, K., Nakamura, R. Browning and insolubilization of ovalbumin by the Maillard reaction with some aldohexaoses. *J. Agric. Food Chem.* 1986, *34*, 351-355.

(5)  Beeley J.G.    The isolation of ovomucoid variants differing in carbohydrate composition. *Biochem. J.* 1971, *123*, 399-405.

(6)  Kato I., Schrode J., Kohr, W.J., Laskowski M.Jr.    Chicken ovomucoid: Determination of its amino acid sequence, determination of the trypsin reactive site, and preparation of all three of its domains. *Biochemistry* 1987, *26*, 193-201.

(7)  Stevens, F.C. & Feeney, R.E.   Chemical modification of avian ovomucoids. *Biochemistry* 1963, *2*, 1346-1352.

(8)  Cho, R.K, Okitani, A., and Kato, H. Polymerization of proteins and impairment of their arginine residues due to intermediate compounds in the Maillard reaction. in *"Amino-carbonyl reactions in food and biological systems. Development in Food Science. 13"* Fujimaki, M., Namiki, M., Kato, H. Eds.    Elsevier, Amusterdam, 1986, pp 439-448.

(9)  Böhlen, O.; Stein, S.; Dairman, W.; Untenfriend, S. Fluorometric assay of proteins in nanogram range. *Arch. Biochem. Biophys.* 1973, *155*, 213-220.

(10) Sakaguchi, S.   A new color reaction of protein and arginine. *J. Biochem.* 1925, *5*, 25-31.

(11) Albanese, A.A.; Irby, V.; Sur, B. The colorimetric estimation of protein in various body fluids. *J. Biol. Chem.* 1946, *166*, 231-237.

(12) Waheed A., Salahuddin A. Isolation and characterization of a variant of ovomucoid. *Biochem. J.* 1975, *47* 139-144.

(13) Kato, Y., Matsuda, T., Kato, N. & Nakamura, R. Maillard reaction of disaccharides with protein: suppressive effect of nonreducing end pyranoside groups on browning and protein polymerization. *J. Agric. Food Chem.*, 1989, 37, 1077-1081.

Chapter 21

# Browning of Amino Acids and Proteins In Vitro: Insights Derived from an Electrophoretic Approach

G. Candiano, G. Pagnan, G. M. Ghiggeri, and R. Gusmano

Nephrology Section, G. Gaslini Institute, 5 Largo G. Gaslini, Genoa 16148, Italy

An interaction of aldoses with aminogroups in aminoacids and proteins takes place "in vitro" under simulated physiological conditions (T=37°C, pH 7.4) which mimic the body environment. The order of reactivity was inversely correlated with the carbon number of the aldose molecule and generated yellow chromophores with characteristic fluorescent spectra. In the case of browning of aminoacids, highly microheterogeneous acidic and basic compounds were generated which were separated by electrophoresis in polyacrylamide gels where yellow pigments presented a different electrophoretic mobility than fluorescent homologs. Titration curves demonstrated a complex behaviour where in all cases pigmented compounds had a cathodic shift for pH<4 or pH>5, while fluorescent homologs migrated to the anode. Browning of albumin by the same aldoses produced numerous microheterogeneus isospecies, with a more anionic pI than the original protein. Finally, highly glycosylated anionic albumin was detected in urines of normal human beings posing the possibility that browning of proteins affecting the isoelectric points of the protein may occour "in vivo".

A chemical interaction between sugars and aminogroups (Maillard reaction) widely occurs in proteins and aminoacids during preparation and storage of food, catalyzed by stressed conditions of the environment (1). Browning compounds generated by a long series of rearrangements and dehydratations are responsible for changes in flavor, color and nutritional value of food and are suspected to play a mutagenic activity as well (2,3). Sugars interact with proteins also "in vivo" under physiologic conditions; the final product still requires a chemical characterization in spite of the fact that browning compounds participate to a series of important pathologies such as diabetic complications (4), a few sequelae of uremia and seem even implicated in aging (5). One of the most serious motif which makes difficult the research on browning compounds is the lack of available techniques for their separation, generally based on liquid chromatography. A still unexplored field is the utilization of electrophoretic techniques based on the assumption that during the process of browning changes in electrical charge may occur in both aminoacids and proteins. However data on this point are lacking. We present here an analysis of the electrical characteristics of browning derivatives occuring from the interactions of sugars, aminoacids and proteins under simulated physiologic conditions (37° C, pH 7.4). Evidence is also presented for the

0097–6156/96/0631–0232$15.00/0

occurrence "in vivo" of changes in electrical charge of selected proteins such as albumin which may bring about some pathogenetic potential.

## EXPERIMENTAL

**Incubations.** Amino acids (40mM) were incubated at 37°C with various monosaccharides (20mM) in the dark under nitrogen in 100mM phosphate buffer pH 7.4. In some experiments, sodium cyanoborohydride (100mM) was added simultaneously with the sugar. Sterilised serum albumin (1mg/mL) purified by pseudoligand chromatography on Affi-gel blue Sepharose (BioRad, Richmond, CA) was incubated in 100mM phosphate buffer (pH 7.4) with 20mM D-glucose, 2-amino-2deoxy-D-glucose, D-ribose, D-erythrose, DL-glyceraldehyde 3-phosphate, DL-glyceraldehyde, and acetaldehyde at 37°C in the dark for up to 7 days in the presence of gentamycin sulphate to prevent bacterial growth. Incorporation of sugars was followed either by the thiobarbituric acid assay (6) after acid hydrolysis of the protein or by monitoring the disappearance of the Lys ε-amino groups with trinitrobenzene-sulfonate (7). After incubation, each sample was ultrafiltered (Amicon PM 30 membranes).

**Electrophoretic titration curves.** Electrophoretic titration curves (8) were performed in polyacrylamide gels (0.75mm thick) supported by silanised glass plates. The polyacrylamide gel slab (12 x 12cm) was cast to contain 5%T, 3%C matrix and 2% (w/v) LKB Ampholines (in the following percentage ratios: 45% pH 3-10, 15% pH 4-6, 15% pH 8-9.5 and 25% pH 9-11). The first dimension (isoelectric focusing of carrier ampholytes) was run at a constant 10W untill the steady state (800V, 11mA) was reached (usually 1h). The second dimension (electrophoresis perpendicular to the pH gradient) was then run at a constant 700V for 30min.. In both dimentions the electrode strips were impregnated with 0.2M sodium hydroxide (cathode) and 0.2M orthophosphoric acid (anode).

**Isoelectric Focusing.** This was performed by using ultrathin-layer (240$\mu$m) polyacrylamide slab gels (5%T,3%C,) cast of silanised glass plates (13x13cm). Gaskets were formed with Parafilm rectangles, one layer giving ~120 $\mu$m thickness.The polymerisation solution contained 12% of glycerol, ammonium persulphate as catalyst, N,N,N',N'-tetramethylenediamine and 2.5% w/v of LKB Ampholines (50% pH4-6, 22% pH5-7,and 25% pH5-8 ranges). The samples were applied to the gel surface with Whatman 3MM paper strips. The pH gradient was evaluated either with 6 pI marker proteins or with an LKB surface glass electrode at 10°C. The runs were carried out at 10°C, using an LKB Multiphor system for a total of 5000V/h. The gels were prefocused for 1h at 500V and 13W. Staining was effected by the photochemical silver method.

**Traditional electrophoersis** Continuos electrophoresis was performed in acrylamide gels (T=4.3%,C=4%) at constant 50mM Tris-borate pH 8.5 buffer.The runs were continued for 2h at 200 Volts at 10°C.

**Determination of molecular weight** A column (1.7x135cm) of BioGel P2 (BioRad) working at room temperature was utilised to evaluate the molecular weight of browning adducts. Elution was achieved with 100mM NaCl of 5ml/h flow rate.
**Spectra.**U.V.visible spectra were recorded with a Beckman DU-8 spectro-photometer at 33°C. Control Lys and α-Boc- and ε-Boc-modified Lys were initially incubated with sugar in $H_2O$ and afterwards placed in buffers of various pH values (4.5-10) Fluorimetric spectra were recorded with a Perkin-Elmer MP44A instrument in standard 10mm quartz cuvettes.

## RESULTS AND DISCUSSION

**Browning of aminoacids**. Glycosylation of α-and ε-aminogroups of aminoacids takes place "in vitro" in conditions which mimic the "in vivo" situation. The first step of the reaction is the formation of a Schiff base followed by an Amadori rearrangement and by a long series of further rearrangements. The order of reactivity we observed was inversely correlated with the carbon in length of the sugar being at least 3 carbons necessary for the initial Amadori rearrangement. On the part of aminoacids the order of reactivity was Lys>Gln-Gly>His-Ala-Trp implying the preponderance of two aminogroups in Lys (α and ε) but not explaining the absence of reaction with some other aminoacids such as Phe, Ser, Tyr and Glu (Tab.I).

**Tab. I:** **Reactivity of aminoacids with aldehydes of different carbon lenght.** The first coloumn shows the order of reactivity of the glyceraldehyde with various aminoacids. The second coloumn shows the reactivity of Lysine with various sugars.

| Aminoacid | Absorbance (340 nm) | Reagents | Absorbance (340 nm) |
|---|---|---|---|
| Ala | ++ | Lysine + D-glucose | - |
| Arg | + - | Lysine + 2-amino-2-deoxy-D-glucose | +++ |
| Asp | - | Lysine + D-ribose | + - |
| Cys | - | Lysine + D-erythrose | +++ |
| Gly | +++ | Lysine + DL-glyceraldehyde-3-phosphate | + |
| Gln | +++ | Lysine + DL-glyceraldehyde | ++++ |
| Glu | - | Lysine + acetaldehyde | - |
| His | ++ | | |
| Lys | ++++ | | |
| Phe | - | | |
| Ser | - | | |
| Tyr | - | | |
| Trp | ++ | | |

The reactivity as a function of the $NH_2$-COOH distance in aminoacids increased almost linearly with a shallow slope up to four carbon atoms and then with a steep slope (Fig.1). All reactive compounds generated brown, intense fluorescent pigments with a characteristic fluorescence spectrum (Ex. 340 nm, Em. 420-430 nm). In two separated approaches, some adducts deriving from Lys and different sugars were subjected to simple electrophoresis in polyacrylamide gels with constant pH 8.5 and a second experiment was performed with equilibrium isoelectric focusing in a narrow range of pH from 4 to 7. In polyacrylamide gels all the adducts of Lys with different sugars behaved as microheterogeneous compounds with different charges migrating towards both the anode and the cathode (Fig.2). Part of these compounds were intensely brown (dark band) while some others were intensely fluorescent (blank band). The most heterogeneous browning derivatives were obtained upon reaction of Lys with glucosamine (Fig.2, track 3) in which case the fluorescent bands predominated over brown compounds. The reaction of Lys with erythrose (track 2) and with glyceraldehyde (track 5) produced instead a lot of pigmented bands with various intensity from dark to mild yellow. Since in these conditions the electrophoretic mobility is only due to charge, it can be concluded that several adducts with different

charge are formed by Lys and sugars of different length. By isoelectric focusing the
adduct of Lys with glyceraldehyde was resolved into at least 8-9 isoelectric species with
pI values in the pH range 4-6. The strongly acidic character of some of these
derivatives suggests the formation of oligomeric products with the presence of 4-7
carboxyl groups counterbalanced by 2-3 protonated aminogroups resulting from
condensation of at least 4-6 Lys into a single molecules. The presence of artifacts due to
aspecific binding to carrier ampholytes were excluded by repeating the experiment in
8M Urea which disrupts such a complex. The analysis of pooled browning adducts of
Lys and glyceraldehyde by gel filtration on Bio-Gel P2 (a chromatographic procedure
which separates molecules on the basis of their molecular mass) revealed the presence
of at least 3 peaks with apparent molecular weight between 1,000 and 500 (9),
indicating policondensation products. Alternatively if we assume that only small
adducts of Lys were formed, additional acidic groups must have been generated in
order to produce isoelectric species in the pH 4-5 range. To further investigated this
aspect we utilized electrophoretic titration curves in which Lys-sugar adducts were
forced to migrate perpendicularly to a quasi stationary pH gradient obtained by
previously focused carrier-ampholytes (8). In this condition Lys-ribose adducts had no
mobility in a range of pH between 4 and 5 while they had both cathodic and anodic
shift at pHs higher and lower than the point of no mobility corresponding to the
isoionic point (Fig. 3A). Interestingly, the isospecies with cathodic shift were intensely
fluorescent (Fig. 3B) while the brown species had the anodic shift. This behaviour is
shown in Fig.3 where it is evident that at least two brown components with
electrophoretic mobility at pH<4 and pH>5 did not appear when the gel was evalutated
with short wave U.V. light ($\lambda max=295nm$).We have no reason to explain the presence
of double anodic shift for pigmented adducts at lower and higher pH than 4-5 and
comparably unexplained is the double catodic shift for fluorescent species. We
tentatively suggest the formation of new chemical groups of unknown structure and
different electrophoretic behaviour for fluorescent and coloured molecules. Further
experimental work is now needed to characterize the chemical structure of browning
products and in this light electrophoretic techniques for separation of different adducts
appear most promising to achieve suitable material for further spectroscopic analysis.
So far we have characterized a single component of the reaction of lysine with
glucosamine as 2,5-Bis-[Tetrahydroxybutyl]-Pirazine (10).

**Browning of proteins.** Human albumin, the most representative of serum proteins,
undergoes a rapid non-enzymatic glycosylation by various sugars during which, after
the formation of a Schiff adduct and an Amadori rearrangement, successive
dehydratation and rearrangement steps take place. The final products are pigmented,
highly fluorescent compounds with the typical optical and fluorimetric features of
brown aminoacids. The analysis of albumin pIs produced by browning revealed a
microheterogeneous spectrum of bands (pI between 4.5 and 5.5) with some unique
characteristics for each aldose derivative (Fig.4). There was in fact a correlation
between the pI of the modified albumin and the molecular weight of the reacting aldose
where low pI species were produced by the shortest sugars (Fig.5). Accordingly, the
most acidic species were produced upon reaction of serum albumin with
glyceraldehyde. On the other hands, the anionization of serum albumin due to any
further rearrangement of the aldose molecule following the initial Schiff base adduct is
demonstrated by the maintainance of a normal pI when the initial glycosylation product
was blocked by sodium cyanoborohydride. Therefore, some interesting similarities
exist between browning of aminoacids and of human albumin "in vitro" as regards to
the reactivity with different aldoses and the formation of new polydispersed populations
of bands with different charge.
When assessing any variation in albumin charge induced by browning one should
consider that at least three relevant phenomena may take place: 1) glycosylation of

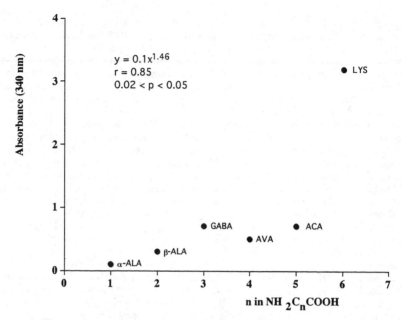

**Figure 1.** Reactivity of glyceraldehyde with amino-derivates of various lenght. The formation of browning derivates (Abs 340 nm) is plotted against the number of carbons in the molecule ($\alpha$-ALA=$\alpha$-alanine; $\beta$-ALA=$\beta$-alanine; GABA=$\gamma$-aminobutyrric acid; AVA=amino valeric acid; ACA=amino caproic acid; Lys=lysine). (Adapted from ref. 9).

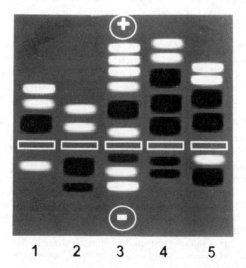

**Figure 2.** Schematic rappresentation of electrophoresis analysis in T= 3.5% polyacrilamide gels of browning adducts of Lys with several aldoses: 1) Lys+glucose; 2) Lys+erythrose; 3) Lys+glucosamine; 4) Lys+ribose; 5) Lys+glyceraldehyde.

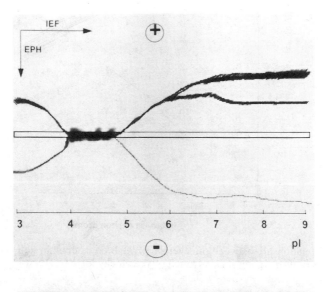

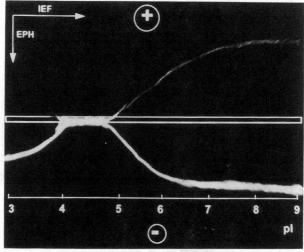

**Figure 3**. Titration curves of the browning derivative of Lys with ribose: A) spontaneous appearance of brown-yellow adducts; B) fluorescent bands under U.V. light (λmax. 295 nm). IEF, isoelectric focusing (first step); EPH, electrophoretic course of the sample (second step).

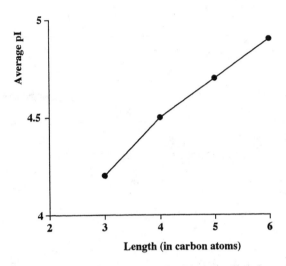

Figure 4. Relationship between the average pI of brown albumin and the length in carbon atoms of the reacting sugars from 3 (glyceraldehyde) to 6 (glucose).

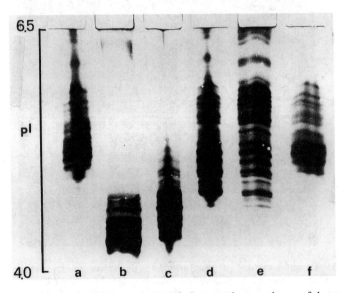

Figure 5. Isoelectric focusing analysis of the reaction products of human serum albumin with various sugars: a) albumin alone; b) albumin+glyceraldehyde; c)albumin+erythrose; d) albumin+ribose; e) albumin+glucosamine; f) albumin+glucose (Adapted from ref. 14).

cationic ε-aminogroups of Lys should decreased "per se" the net surface charge of the protein; 2) a change in protein conformation may exposed aminoacids with charged groups otherwise buried in hydrophobic domains; 3) new charged groups can be generated by reaction of aldoses with aminogroups, in a way which recalls the simple reaction of aldoses with aminoacids. Data are available in the literature suggesting the contribution of all these three pathways in producing changes in albumin pI. First of all, it has been already demonstrated that the process of glycosylation of proteins is massive and produces new adducts where the molar ratio of aldose versus Lys is greater than 3, suggesting saturation of most of the exposed ε-aminogroups. Changes in conformational properties of albumin have been demonstrated by circular dicroic spectra (*11*), indicative of modest albeit significant variations in α-elix content. Changes in conformation may support the implication of the exposure of otherwise bured charged groups. In the case of albumin, for example, there are 59 Lys residues among the 575 aminoacids of the whole sequence but only 16 are exposed on the surface owing to a complex secondary and tertiary structure of the protein. The exposure of some of these bured ε-aminogroups should introduce new charges on the net overall equilibrium. Finally, the formation of new charged groups on the surface of the proteins may be inferred from the analysis herein presented of the electrical charge of browning aminoacids.

**Browning of proteins "in vivo".** It is now generally accepted that non enzymatic glycosylation of ε-aminogroups of several proteins takes place in healthy humans and that this reaction is accelerated in diabetic patients. Evidence is available that browning compounds are involved in the pathogenesis of diabetic sequelae including microangiopathy, cataract and diabetic nephropathy in both human beings (*5*) and animals (*12*). The discussion about an implication of browning compounds in the pathogenesis of diabetic sequelae is besides the scope of this paper, however a point is noteworthy here. This refers to the detection of the electrical charge of glycosylated albumin "in vivo" with references for its metabolic fate. In 1985, Ghiggeri and coworkers (*13*) were able to show the presence of widely microheterogeneus isoforms of albumin in urine of normal human beings, with an anionic charge compared to the serum homologue (Fig.6). The most anionic albumin isoforms were also identified as the most glycosylated by affinity chromatography on Concanavalin-A Sepharose and by carbohydrate analysis. Although no information is available on a possible identification of anionic isoforms as browning derivates, based on what shown by "in vitro" studies on the electrical charge of browning albumin this possibility seems reasonable. It is interesting that the most anionic and glycosylated albumin isoforms are massive escreted into urines which means manipulation by the kidney, one of the targed organ of the diabetic milieu.

**Conclusive remarks.** It seems therefore conceivable that following non-enzymatic glycosylation and successive rearrangements, aminoacids and proteins undergo a profound structural transformation involving the formation of new chemical structures. One important consequence is the generation of microheterogeneus products each bearing different electrical charge and spectroscopic characteristics. Based on the difference in charge, electrophoretic techniques may be utilised to separate and characterize different browning compounds. The occurrence in human beings of glycosylated compounds with similarities in electrical charge with the "in vitro" products of browning and their demonstrated roles in the pathogenesis of several diseases make any effort in this are worthy to be done.

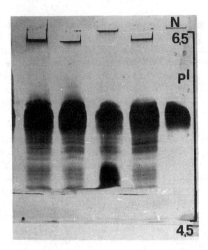

**Figure 6.** Isoelectric focusing of urinary albumin purified with pseudo-ligand chromatography on Affi-gel Blue Sepharose from normal human beings. Numerous bands are visible with a pH more anionic compared to the unmodified homolog (N). (Adapted from ref. 13).

## Literature Cited

1)  Ledl, F. In *The Maillard reaction in food processing, human nutrition and physiology* ; Eds. Finot, P.A., Aeschbacher, H.U., Hurrel, R.F., Liardon, R.; Birkhäuser Verlag: Basel, Boston, Berlin, 1990, pp 19-42.
2)  Reineccius, G.A. In *The Maillard reaction in food processing, human nutrition and physiology* ; Eds. Finot, P.A., Aeschbacher, H.U., Hurrel, R.F., Liardon,R.;Birkhäuser Verlag: Basel, Boston, Berlin, 1990, pp 157-170
3)  Hurrel, R.F. In *The Maillard reaction in food processing, human nutrition and physiology* ; Eds. Finot, P.A., Aeschbacher, H.U., Hurrel, R.F., Liardon, R.; Birkhäuser Verlag: Basel, Boston, Berlin, 1990, pp 245-258.
4)  Vlassara, H., Bucala, R., Striker, L. *Lab.Invest.* **1994**, 70, 138-151.
5)  Beisswenger, P.J., Moore, L.L., Brink-Johnson, T., Curphey, T. *J. Clin. Invest.* **1993**, 92, 212-217.
6)  Dolhofer, R., Wieland, OH.*FEBS Lett.* **1979**, 103,282-286.
7)  Eklund, A. *Anal. Biochem.* **1976**, 70, 434-439.
8)  Valentini, L., Gianazza, E., Righetti, P.G. *J. Biochem. Biophys. Methods* **1980**, 3, 323-328.
9)  Candiano, G., Ghiggeri, G.M., Delfino, G., Queirolo, C., Cuniberti, C., Gianazza, E., Righetti, P.G. *Carbohydr. Res.* **1985**, 145, 99-112.
10) Candiano, G., Ghiggeri, G.M., Gusmano, R., Zetta, L., Benfenati, E., Icardi, G. *Carbohydr. Res.* **1988**, 184, 67-75.
11) Ghiggeri, G.M., Candiano, G., Delfino, G., Queirolo, C., Vecchio, P., Gianazza, E., Righetti, P.G. *Carbohydr. Res.* **1985**, 145, 113-122.
12) Cohen, M.P., Mud, E., Wu, U.Y. *Kidney Int.* **1994**, 45, 1673-1679.
13) Ghiggeri, G.M., Candiano, G., Delfino, G., Queirolo, C. *Kidney Int.* **1985**, 28, 168-177.
14) Candiano, G., Ghiggeri, G.M., Delfino, G., Queirolo, C., Gianazza, E., Righetti, P.G. *Electrophoresis* **1985**, 6, 118-123.

Chapter 22

# Instrumental Means of Monitoring the Flavor Quality of Foods

G. A. Reineccius

Department of Food Science and Nutrition, University of Minnesota, 1334 Eckles Avenue, St. Paul, MN 55108

It has been a long term goal of many researchers to use instrumental means to replace some sensory functions. The use of sensory panels for quality control purposes presents many problems which may be minimized through the use of supplementary instrumental techniques. Over the years, gas chromatography and mass spectrometry have found limited application for this purpose. Recently an instrument generically called an "electronic nose" has been commercialized. This paper will present a brief overview of gas chromatographic and mass spectral techniques used to monitor flavor quality in foods but focus on the new electronic nose instruments.

It is extremely doubtful that there will be any reliable instrumental method in the foreseeable future that will replace the human being for the sensory evaluation of foods. However, there are various instrumental techniques that can be used to supplement sensory analysis (1). These techniques are typically simple rapid screening procedures which reduce the burden on sensory analysis but do not eliminate it. As an example, we see very common usage of headspace gas chromatography to monitor hexanal in vegetable oils. There is a well established correlation between the oxidation of vegetable oils and hexanal.

I will present a discussion of some of the instrumental techniques in use today for monitoring the flavor quality of foods. I am going to limit this discussion to the aroma portion of flavor and ignore the taste of foods. This is not meant to imply that taste is not important to flavor. However, the instrumental methods used to monitor aroma and taste are very different and justify separate discussions. Also taste is often created by the food company through formulation and thus is easier to monitor since it may involve measuring sugar concentration or pH. However, when the taste is the result of (or modified by) the food processing operation (e.g. coffee roasting), monitoring taste quality is an extremely difficult task. Unfortunately, taste is contributed by semi or nonvolatiles in foods and this class of flavoring materials

0097–6156/96/0631–0241$15.00/0
© 1996 American Chemical Society

is only poorly understood. Thus, analytical methods have not been well developed (2). Some pioneering work in this field was reported at an ASIC meeting (3) where lipid-based sensors were being used to determine the taste properties of coffee.

In the following paper, I have included a limited discussion of the traditional methods of monitoring aroma quality of foods such as gas chromatography and mass spectrometry but have focused to a greater extent on the new "electronic nose" methodologies.

## Gas Chromatography/Mass Spectrometry

**Aroma Isolation.** The initial task in this approach to monitoring flavor quality is to deliver a sample of aroma to the instrument. This may appear to be simple task but it is generally very complicated and determines the subsequent success of the analysis. Our major problem is that our instruments are not nearly as sensitive as the human nose so generally some method of concentration must be included in the analysis. This requirement may be simplified by the fact that the isolation method may not have to be all inclusive. For quality control we simply need to find indicators of flavor quality rather than isolate all of the compounds which may contribute to flavor. However, the method must be simple, rapid and reproducible.

The flavor isolation methods which most readily lend themselves to quality control applications are static headspace, dynamic headspace, direct injection and solvent extraction techniques. Since there are numerous recent reviews in the literature on these methods, there is little need to present any detail here but only summarize the key points about a given method. The reader can refer to reviews provided by Jennings and Shibamoto (4); Reineccius and Anandaraman (5), Reineccius (6) or Teranishi and Kint (7) for more detail.

Headspace techniques are often the method of choice since there is virtually no sample preparation involved. One simply places the food sample in a closed vessel, allows the headspace to equilibrate and then samples the headspace with a gas-tight syringe or an automated sampling system. Problems with sample carry over in the syringe and reproducibility favor automated systems for headspace sampling, The primary limitation of headspace sampling is a lack of sensitivity. One may not isolate sufficient quantities of indicator compounds to permit accurate and precise quantification. However, the simplicity, reproducibility and speed of this method make it exceptionally desirable for quality control purposes.

Headspace concentration techniques have some application in the quality control area since they offer a means of delivering more material to the instrument. In headspace concentration techniques, the sample is purged with an inert gas (perhaps liters of headspace), the volatiles are passed through some type of trapping system and then the concentrated volatiles are stripped from the trap for introduction into an analytical system. The trap traditionally has been filled with a porous polymer (e.g. Tenax®), which after loading with volatiles may be thermally desorbed or solvent extracted (e.g. with diethyl ether). Thermal desorption is quite slow (up to 3 min.) and results in poor chromatography of the early eluting compounds unless some cryofocusing technique is used. Solvent extraction of the Tenax® and then liquid injection of the elution solvent has become popular in research studies. However, in the case of quality control, thermal desorption has an advantage in terms of time and would be the preferred approach. Other trapping methods are employed and discussions of them can be found in many of the references cited earlier.

Direct injection techniques also may be used for this purpose. Many food products are liquids (either fat or water-based) which can be readily sampled by automated liquid sampling systems. Sensitivity also is often quite adequate, If one considers compounds present in a food at concentrations > 1 ppm (i.e. 1 $\mu g/g$), there would be > 1 $\eta g/mg$ of those compounds in the food. A 20 $\mu L$ injection would provide 20 ng to the gas chromatograph. That is very adequate for detection and accurate integration. The obvious problem here is sample decomposition in the heated injection port of the gas chromatograph, thereby producing artifacts. This problem has been addressed in various ways as is discussed in the literature and thus the approach can be used for delivering aroma to the instruments for further analysis.

Solvent extraction techniques may be applied to foods which contain no fat. Fat would be extracted along with the flavor compounds, thereby limiting concentration and confounding chromatography. Therefore, solvent extraction techniques are limited to foods such as fruits, fruit juices, essences, wines, etc. which are fat free.

**Aroma Analysis.** Once we have obtained a suitable aroma isolate, we must turn to its' analysis. Here we are concerned with the separation and quantification of individual flavor compounds from complex mixtures. The most commonly used method for flavor analysis is gas chromatography. Owing to the complexity of the separations, we are virtually always working with high-resolution capillary chromatography. Mass spectrometry also finds substantial use in the analysis of aromas. Mass spectrometry is used or, unfortunately, misused too often. Mass spectrometry should be used in quality control situations when it will speed analysis, improve sensitivity, or provide selectivity otherwise unavailable. All too often, the mass spectrometer operator is in research and has a strong desire to identify everything in the gas chromatographic run. In fact, it is not necessary to know the identity of any flavor compounds used in predicting or monitoring flavor quality. One typically draws some correlation between flavor quality and the concentration of some gas chromatographic (or mass spectral) peaks. While it may be esthetically satisfying to identify the indicator peaks, it is most often not necessary and may be a waste of time.

As noted, gas chromatography is most commonly used in the analysis of food aromas. However, the value of mass spectrometry (MS) in reducing analysis time has been nicely demonstrated by Labows and Shushan (*8*). In this particular study, they used MS/MS to analyze the volatile constituents in knockwurst sausage. This analysis by gas chromatography (GC) required headspace concentration followed by gas chromatography. While the authors did not state the time required for the total gas chromatographic procedure, it would most likely be about an hour. In comparison, the MS/MS procedure took less than 1 min. For the MS/MS procedure, gas was passed over the sausage and then directly into the ion source of the MS/MS. There ionization occurred and an ion chromatogram was generated. This ion chromatogram may suffice in many cases to provide the necessary quantitative data and the analysis time is only a few seconds.

**Data Analysis.**   Whether we have obtained an analytical profile of the food aroma by gas chromatography or mass spectrometry, it will be necessary to do some type of data analysis. This analysis can be extremely simple such as quantifying a single indicator peak in the chromatogram (e.g. hexanal) or it may be very complex involving multivariate statistics of the entire chromatographic output as well as sensory panel data.

The chromatographic peaks analyzed and ultimately used as predictors of flavor quality may be responsible for the aroma quality or they may simply be correlated to the causative compounds. While one would feel more comfortable monitoring the compounds responsible for a particular sensory attribute, it is not necessary in order to accomplish the task at hand. The example noted earlier of monitoring the flavor quality of vegetable oils is such as situation. Most often, the analyst focuses on the presence of hexanal as an indicator of oxidized off-flavor when, in fact, the unsaturated aldehydes may play a more important role in determining sensory quality.   Since the hexanal is present in greater quantity than the unsaturated aldehydes, it is easier to accurately monitor the hexanal and thus predict sensory quality.

Other examples of using a single compound to predict flavor quality include flavor defects in dehydrated potatoes (9), milk (10), and cheese (11).  It is also possible to use total volatiles to predict flavor quality, As an example, Dupuy et al, (12) have used total volatiles to predict flavor score in vegetable oils. Since flavor differences very often are the result of differences in a number of volatile constituents rather than one or two individual chemicals, the best sensory prediction is often obtained using several gas chromatographic peaks. The selection of these peaks and development of the proper mathematical equations (models) for predicting flavor quality involves the use of multivariate statistics. Comprehensive reviews have been provided by Powers (13) and Powers and Moskowitz (14) on the application of multivariate statistics to the correlation of instrumental and sensory data.  A text which covers this subject in substantial detail has been authored by Sharaf et al. (15) and a significant chapter offered by Leland and Lahiff (16). Most of the past work has used either multiple regression or discriminant analysis.  The goal of multiple regression is to model a continuous response variable as a function of two or more predictors.  For discriminant analysis, the goal is to put a sample into a class such as degree of off flavor or geographical origin.

**Multiple Regression Analysis.**   Through the application of multivariate regression, several flavor problems have been approached.  For example, Pino (17) determined a relationship between orange juice volatiles and sensory panel preference scores.  Galetto and Bednarczyk (18) used similar techniques in the analytical prediction of overall onion flavor (based on three gas chromatographic peaks).  Manning (19) used multiple regression to correlate taste panel evaluations of cheddar cheese flavor to headspace volatiles.   Recently, wine has been characterized by such methodology (*20*).  While regression analysis is most properly applied to continuous response variables, it has also been applied to Hedonic scores.

**Discriminant Analysis.** Discriminant analysis is used to model a categorical response to a variable, for example, a flavor or treatment grouping, as a linear function of two or more predictors.  Powers and Keith (*21*) published one of the earliest papers on the use of stepwise discriminant analysis (SDA) for gas chromatographic data.  They were able to classify coffees using this technique. Other applications include work on wine classification (*22*) and sweet potato classification (*23*).

In some cases, both multiple regression and discriminant analysis have been applied to flavor problems.  Examples include the work of Liardon et al. (*24*) on coffee, Leland et al. (*25*) on milk off-flavors and Aishima (*26*) on soy sauces.

**Limitations.** Irrespective of the statistical method selected, the goal must be to determine which peaks in the gas chromatographic profile are the best indicators of sensory quality.  This peak selection together with a mathematical model are used to analytically monitor (or predict) sensory quality.  Assuming that the analytical method is both accurate and precise, an instrumental method such as this can very effectively predict sensory quality.  In a quality control laboratory, this analytical approach can serve as an untiring evaluator.

We do, unfortunately, run into problems with this approach.  The main problem is the need to deliver an appropriate, adequate and reproducible amount of aroma isolate to the instrument.  Even the simplest sampling systems can be a problem on a daily basis.  A secondary problem relates to the analytical system.  It must accurately and reproducibly provide quantitative data on some constituent(s) in this aroma isolate which permits the sensory prediction.  These requirements are formidable and introduce additional elements of variability.  Finally, one must acknowledge that the operator of such a system must be highly trained and knowledgeable in gas chromatography and/or mass spectrometry.  The operator also may have to be familiar with multivariate statistics and sensory analysis to set up the initial correlations to permit sensory predictions.  These requirements have resulted in scientists searching for simpler means of accomplishing the task of using instruments to provide sensory predictions of food flavor quality.  The following section of this presentation discusses a relatively new approach, the use of an electronic nose, to achieve instrumental predictions of sensory quality.

**Electronic Nose**

The initial technology for the electronic nose came from the US Air Force research program on the Stealth bomber (*27*).  They were interested in developing polymers which would conduct electricity and thus be useful in evading enemy radar.  When the military gave up on this approach, there was enough information published in the literature to interest scientists in applying the technology to electronic noses.

The electronic nose has been in development more than 10 years at Toulouse (France), Warwick (UK) and Southampton (UK) Universities (*28*).  As a result of this effort, three companies have offered commercial instruments for sale (Alpha M.O.S., AromaScan and Neotronics). The primary differences between these manufacturers is the type and means of manufacturing the aroma sensors.

Competing methods of monitoring food aroma (gas chromatography and mass spectrometry) are based on the principal of separating the aroma of food into individual components and then measuring them. The human olfactory system does not effect a separation but takes the aroma as a complex mixture and subjects it to a sensor array. The human olfactory system (sensor array) then collectively responds to the aroma and sends a complex signal to the brain which does pattern recognition and arrives at a decision about the aroma. For example, the brain will judge that the food just smelled is banana flavored or perhaps it is a good cup of coffee. The individual olfactory receptors are not particularly selective but differ primarily in sensitivity to a given odorant. This is demonstrated in the neural response pattern of the frog to geraniol (or camphor) that is presented in Figure 1.

The electronic nose is designed to mimic this operating principal. The electronic nose uses from 6 to 32 different sensors aligned in an array to respond to food aroma. The food aroma is passed across the sensor array and the array output is sent to a computer that makes a correlation between this sensor pattern and that of a sensory panel. Then employing classical multivariate statistics (pattern recognition programs) or artificial neural network algorithms, the instrument can, with training, give a classification of an unknown sample. The sample analysis time will generally range from 2-8 min. This includes the time required to introduce the sample, obtain a sensor response, analyze the data and regenerate the sensors . Training of the instrument is done using a set of 23 samples (28). This is considered adequate to account for variability in samples, sampling techniques and instrument response.

**Sensors.**   As was mentioned earlier, the primary difference between manufacturers is the type and means of making the sensors. At the present, the Alpha M.O.S. relies primarily on metal oxide sensors while both AromaScan and Neotronics use polymer-based sensors. Both of these types of sensors will carry an electrical current which changes in the presence of a volatile substance (the aroma of food for example). This change in electrical is then used as the sensor output or response to an aroma.

The metal oxide sensors are typically based on tin oxides with varying amounts of the catalytic doping metal (Pd or Pt) (29). However, sensors can also be made with other metals (30). Metal oxides are semiconducting materials which are gas sensitive. Oxygen in the air reacts with lattice oxygen vacancies in the bulk material removing electrons from the material.

$$n + 1/2 O_2 ----> O_{(s)}^-$$

While the oxidation state is temperature dependent, the O negative species occurs at about 400C. In the presence of an aroma molecule, the chemisorbed oxygen species reacts irreversibly and reaction products are formed (typically $CO_2$ and water).

$$R(g) + O_{(s)} - ----> RO(g) + n$$

The resistance of the sensor is thus decreased, the magnitude being dependent upon the type of sensor and the aroma molecule. This change in resistance is the sensor output sent to the computer. The response time of the sensor depends on the reaction kinetics, the headspace and the volume of the measured headspace. This generally takes about 10 to 120 sec (*29*). Metal oxide sensors are claimed to offer good sensitivities to a broad range of organic molecules (ppm to ppb).

The polymer-based sensors are typically made by depositing polypyrrole resins (semiconducting materials) across two electrodes (*28*). The polymers made by Neotronics are electrochemically produced while those of AromaScan are "inked" or masked". These resins have an inherent electrical resistance which changes as organic molecules are absorbed. They typically will respond to molecules ranging in molecular weight from 30 to 300 which covers the range normally associated with aroma compounds. They give a strong response to molecules which such as alcohols, ketones, fatty acids and esters. They will give a reduced response to fully oxidized species such as $CO_2$, $NO_2$ and $H_2O$. Also, they are particularly sensitive to aroma molecules containing sulfur or amine groups. The extreme sensitivity of the nose to similar compounds is well known.

**Data Analysis.** While the data can be processed in a variety of ways employing classical multivariate statistical methods (e.g. discriminant functions, K-nearest neighbor, template matching, cluster analysis and partial least squares), the most simple method from a user viewpoint but powerful is the artificial neural network approach (ANN). ANN systems are self learning - the computer is given the sensor array input and the sensory panel data and after several samples, establishes a mathematical equation involving the sensor array input and sensory panel data that best predicts the sensory panel judgment. One can see that the greater the amount of data given the instrument the better the predictions become. The system can be trained to give responses about odor notes (e.g. pear-like, green or buttery), quality (good or bad) or acceptance (accept or reject). Odors that have not been given to the instrument or fit the established training profile will be judged as being unidentified. The sensory profile of these samples can be given to the computer and these new odors then will also be categorized. Thus the system is always able to be further trained.

**Applications:**  While few applications are available in the scientific literature for the AromaScan and Neotronics instruments, numerous applications of the Alpha M.O.S. system occur in the literature. The breadth of the applications is impressive. The Wall Street Journal (*27*) reported that the electronic nose was being evaluated for possible applications including: new car smell (General Motors), deodorants (Unilever), perfume creation (possible patent implications), wines (Wine Magazine), breath (indication of diabetes), infection of wounds (South Manchester University Hospital), sewage treatment plants, fish freshness (FDA) and numerous others. A few of the food related applications follow (Figures 2-6). Most of these applications are self explanatory so little will be said about them.

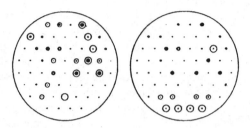

Figure 1. Representation of the unitary activity of the olfactory nerve of a frog in the presence of geraniol or camphor (*32*)

Columbian                                     Brazilian

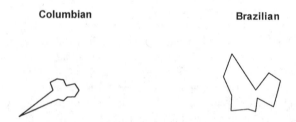

Figure 2. Signatures of two coffees (Brazilian on the left, Colombian on the right) using an array of twelve sensors (*32)*.

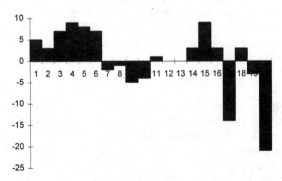

Figure 3. Difference plots of fresh and rancid mayonnaise (*31*).

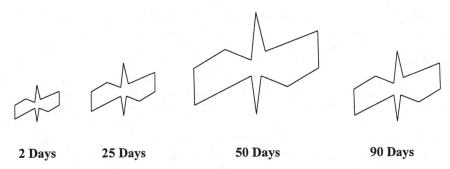

**2 Days**          **25 Days**              **50 Days**                **90 Days**

Figure 4. The volatile profile above fermented sausages during ripening (*29*).

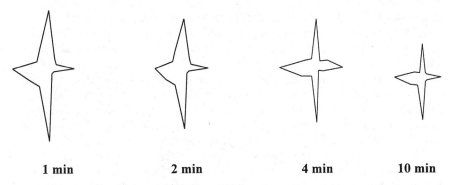

**1 min**              **2 min**              **4 min**              **10 min**

Figure 5. The volatile profile above chewing gum after various chewing times (*29*).

**Gin 1**                                        **Gin 2**

Figure 6. The volatile profile above two different gins (*29*).

## Summary

Gas chromatography has found use as a means of monitoring the flavor of foods in a large number of research applications but a limited number of non research applications. Transferring the method from the research laboratory to a quality control situation has proven difficult. When it has been possible to prepare an aroma isolate in a limited time (e.g. static headspace) and a single peak has proven adequate to predict the aroma of a food, gas chromatography has been quite successful. The stumbling block has proven to be time limitations or complexity of the method. Mass spectral methods have proven about equally limited. Mass spectrometry offers time savings in some applications which has proven valuable. However, the cost and additional complexity have also proven to be a barrier to wider use of this approach.

The appearance of commercial systems employing "electronic nose" methodology is most enticing. They do not require isolation of the food aroma but are designed to directly sample the headspace of the food. Additionally, they do not require any separation of the aroma. Both of these considerations result in decreased analysis time. They also offer Artificial Neural Network data analysis. This means that the data analysis system itself is capable of establishing the required mathematical relationships between sensory data and instrumental data. The system optimizes its ability to discriminate between samples and this does not have to be done by the operator or outside data analysis. Commercial electronic nose systems essentially are "turn key" systems requiring little operator skill. It appears that these systems circumvent many of the problems that limit the application of gas chromatography and mass spectrometry for the monitoring of food aroma.

However, before one becomes too excited about the electronic noses, there are several questions that need to be investigated. A major concern is for the long term stability of the sensor array. If the instrument is to be used for the quality control of food aroma, the sensors must be stable over time. Otherwise, the instrument will have to be frequently calibrated which will consume significant sensory panel time. There is also concern for the "poisoning" of the sensors. It is anticipated that high levels of some volatiles (e.g. acids) may poison the sensors and radically change their response. Additionally. one must consider whether the sensors will be sensitive enough to detect many of the volatiles responsible for desirable and undesirable aromas in foods. There may well be situations when a food becomes contaminated with compounds such as chlorophenols or geosmin. It is highly unlikely that the electronic noses will have the sensitivity to detect these compounds. The argument may be that these compounds will likely not occur alone but with other compounds (e.g. microbial metabolites) which may be present at higher levels and be detected by the electronic nose. It is understood that the sensors must only detect compounds that reliably indicate a desirable or undesirable aroma - they do not have to actually detect causative aroma compounds.

As to the ability of electronic noses to perform quality measurements of food aroma, there will be no solution other than trying it for a particular application. They may well revolutionize the way quality control is done on foods.

## References

*1).* Dirinick, P.; DeWinne, A. Advantages of instrumental procedures for the measurement of flavour characters. In: *Trends in Flavor Research;* H.Maarse and D.G. van der Heij, eds; Elsevier Publ.: Amsterdam, 1994 p. 259.

*2).* Pickenhagen, W.; Spanier, A. Contribution of low and nonvolatile material to the flavor of foods. American Chemical Society Annual Meeting, Chicago, August 1995.

*3).* Komai, H. Measurement of coffee taste using lipid membrane taste sensors. *Proceedings of the ASIC meeting*, Kyoto, April 1995.

*4).* Jennings, W. G.; Shibamoto, T. *Qualitative Analysis of Flavor and Fragrance Volatiles by Glass Capillary Gas Chromatography.* New York: Academic Press, 1980.

*5).* Reineccius, G. A.; Anandaraman, S. Analysis of volatile flavors in foods. In: *Food Constituents and Food Residues: Their Chromatographic Determination*; Lawrence, J. E. ed. Marcel Dekker: New York, 1984.

*6).* Reineccius, G. A. Isolation of food flavors. In: *Flavor Chemistry of Lipid Foods*; D. B. and T. B Smouse, eds. Amer. Oil Chemists Society: Champaign, Illinois, 1989.

*7).* Teranishi, R.; Kint, S. Sample preparation. In: *Flavor Science;* T.E Acree and R. Teranishi, eds. ACS: Washington D.C., 1994, p 13 7.

*8).* Labows, J. N.; B. Shushan. Direct analysis of food aromas. *Amer. Lab.*, **1983,** 15(3):56.

*9).* Boggs, M.M.; Buttery, R.G.; Venstrom, D.W.; Belote, M.L. Relation of hexanal in vapor above stored potato granules to subjective flavor estimates. *J. Food Sci.* **1964,** 29:487

*10).* Arnold, R.C.; Libbey, L.M.; Day, E.A. Identification of components in the stale flavor fraction of sterilized concentrated milk. *J. Food Sci.,* **1966,** 31(4):566.

*11).* Badings, H. T.; Stadhouders, J.; Van Duin, H. Phenolic flavor in cheese. *J. Dairy Sci.,* **1968,** 51:31

*12).* Dupuy, H.P.; Fore, S.P.; Goldblatt; L.A. Elution and analysis of volatiles in vegetable oils by gas chromatography. *J. Am. Oil Chem. Soc.,* **1971,** 48:876.

*13).* Powers, J.J. Techniques of Analysis of Flavours-Integration of Sensor: and Instrumental Methods. In: *Developments in Food Science. Food Flavours.* Morton, I.D., and A. J. MacLeod, eds., Elsevier: New York, 1982.

*14).* Powers, J.J.; Moskowitz, H.R. Correlating Sensory Objective, Measurements-New Methods for Answering Old Problems. *Standard Technical Publ. 594.* ASTM: Philadelphia, 1976.

*15).* Sharaf, M. A.; Illman, D.L.; Kowalski, B.R. *Chemometrics.* Wiley: New York, 1986.

*16).* Leland, J.; Lahiff, M. Sensory Instrumental correlations in foods. In: *The Source Book of Flavors.* G.A. Reineccius, ed., Chapman and Hall: New York, 1994.

*17).* Pino, J. Correlation between sensory and gas chromatographic measurements on orange volatiles. *Acta Aliment.* **1982,** 11(L):1.

252 CHEMICAL MARKERS FOR PROCESSED AND STORED FOODS

*18*). Galetto, W.G.; Bednarczyk, A.A. Relative flavor contribution of individual volatile components of the oil of onion (*Allium Cepa*). *J. Food Sci.,* **1975**, 40:1165.
*19*). Manning, D.J. Cheddar cheese flavour studies. II. Relative flavor contributions of individual volatile components. *J. Dairy Res.* **1979**, 46(3):523.
*20*). Guedes de Pinho, P.; Bertrand, A.; Alvarez, P. Wine characterization by multivariate statistical analysis of the sensory and chemical data. In: *Trends in Flavor Research*, H. Maarse and D.G. van der Heij, eds. Elsevier Publ. :Amsterdam, **1994**. p. 229.
*21*). Powers, J.L; Keith, E.S. Stepwise discriminant analysis of gas chromatographic data as an aid in classifying the flavour quality of foods. *J Food Sci.* **1968**, 33:207.
*22*). Nobel, A. C.; Flath, R.A.; Forrey, R.R. Wine headspace analysis: Reproducibility and application to varietal classification. *J. Agric. Food Chem.* **1980**, 28(2):346.
*23*). Tiu, C. S.; Purcell, A.E.; Collins, W.W. Contribution of some volatile compounds to sweet potato aroma. *J. Agric. Food Chem.*, **1985**, 33(2):223.
*24*). Liardon, R.; Ott, U.; Daget, N. Application of multivariate statistics for the classification of coffee headspace profiles. *Lebensmitt.-Wissensch. Technol.*, **1984**, 17(1):32.
*25*). Leland, J. V.; Lahiff, M.; Reineccius, G.A. Predicting intensities of milk off-flavors by multivariate analysis of gas chromatographic data. In: *Flavor Science and Technology;* Martens, M.; G. A. Dalen; H. Russwurm, eds. Wiley: New York, 1987.
*26*). Aishima, T. Discriminant and cluster analysis of soy sauce GLC profiles. *J. Food Sci.*, **1982**, 47(5):1562.
*27*). Pope, K. Technology improves on the nose as scientists try to mimic smell. Wall Street Journal March 1, **1995**, pg. B1.
*28*). Marsili, R. The electronic nose: A sensory evaluation tool. *Food Product Design*, **1994**, (June), 53-67.
*29*). Alpha M.O.S. Inc., 1994. Product application information, P.O. Box 459, DeMotte, IN 46310.
*30*). Bartlett, P.N.; Gardner, J.W. Odour sensors for an electronic nose. NATO ASI Series, *Sensors and Sensory Systems for an Electronic Nose*, Reykjavik, Island, **1991**, p. 31.
*31*). AromaScan Inc., 14 Clinton Dr., Hollis, NH 03049.
*32*). Moy, L.; T. Tan, Monitoring the stability of perfume and body odors with the "electronic nose". *Perfum. Flavorist,* **1995**, 19:11.

# Chapter 23

# Correlation Between Color Machine Vision and Colorimeter for Food Applications

**P. P. Ling[1], V. N. Ruzhitsky[1], A. N. Kapanidis[2], and Tung-Ching Lee[2,3]**

**[1]Department of Bioresource Engineering, and [2]Department of Food Science, Center for Advanced Food Technology, Cook College, Rutgers, The State University of New Jersey, P.O. Box 231, New Brunswick, NJ 08903–0231**

Color is an important food quality attribute and its evaluation is critical for food manufacturing, processing and storage. Conventional objective color measurements include colorimetry and spectrophotometry which have limited color sensing capabilities due to low spatial resolution. In contrast, color machine vision (CMV) offers a significantly higher spatial resolution that creates new opportunities for color quality control. We have demonstrated the feasibility of using CMV for color measurement of foods. The color image acquisition system was designed based on a performance model and evaluated against a Minolta Chroma Meter using several standard Macbeth color plates as well as processed beef and carrot samples. Excellent correlation existed for all systems studied (R>0.98). Strong correlations existed between measured external colors and pigment concentrations and the extent of processing in the food products (samples) used in this study. These results suggest that CMV can be used for both off-line and on-line color quality control in food industry.

Color is an important attribute of food quality (Clydesdale, 1993; Bourne, 1982). Consumers select individual food products such as fruits and vegetables primarily on appearance attributes such as color, shape, size and surface defects (IFT, 1990). Of these, color may be the most influential, because off-color food is likely to be rejected even when it has good flavor or texture. Conversely, high color quality food products usually have higher market value. Therefore, the development of food with an attractive appearance is an important goal in the food industry. How to sense

---

[3]Corresponding author

0097–6156/96/0631–0253$16.50/0

color and estimate pigment contents of food materials to obtain desired colors is a challenge. A sensing technology that can rapidly monitor color quality during processing for value-added food products is needed.

Color quality of food can be determined by human inspectors. Although sensory evaluation is appropriate and may be the most accurate method for specifying acceptable color levels of food materials, sensory panels are not always available for routine large scale color quality determination. Moreover, their evaluations are often limited to qualitative descriptions. Human inspection is subjective, with color readings affected by the lighting conditions at the inspection site, the inspector's health and mood.

Objective color measurements have been performed in laboratories using colorimeters and spectrophotometers. Due to their low spatial resolution, conventional instruments are impractical for collecting large number of samples. Because only a few samples can be collected to represent the product, the approach frequently mis-represents food products that have non-uniform color distributions.

Colorimeters also do not measure what humans see in colors of objects. Human color perception is influenced by (1) color quality of light source, (2) background color, and (3) surface texture of the sample. Colorimeter measurements, on the other hand, are performed in tightly controlled illumination chambers and can only collect averaged spectral responses of the sampled area. It is not possible for colorimeters to express human perceived colors that are affected by various light sources or background colors. Surface texture information is also unavailable to colorimeters due to their low spatial resolution.

Machine vision has a significantly higher spatial resolution than that of a colorimeter or a spectrophotometer. Colorimeter can only sample a small area at a time. A common commercial Red, Green, Blue (RGB) CCD camera has more than 240,000 sensing elements and can be used to collect color information from more than 80,000 locations in the field of view. In contrast to conventional color measurement, color machine vision (CMV) offers a tremendous amount of spatial information, which can be used to quantitatively evaluate color distribution. Thus, background color can be identified and surface texture estimated from the color distribution on the surface. Temporal color changes in food can also be quantified by comparing a sequence of machine vision images taken at different times. Machine vision is an ideal sensing tool for monitoring dynamic color changes of food materials and products during storage or other processes.

Machine vision technology has been used for non-contact, non-destructive sensing mostly of industrial parts and limited food products. Industrial applications are widespread in electronics and auto industries. For the food industry, machine vision activities occur mainly in laboratory research with only a few successful industrial applications (Tillett, 1991; Chen and Sun, 1991). These include preliminary work on: (1) cheese shred evaluation (Ni and Gunasekaran, 1994); (2) discrimination of whole from broken corn kernels (Zayas et al., 1990); (3) analysis of kiwi fruit slices (Roudot, 1989); (4) characterization of extrudates (Zhang et al., 1994; Barrett and Peleg, 1992; Alahakoon et al., 1991; Smolarz et al., 1989); (5) determination of doneness of beef steaks (Unklesbay et al., 1986); (6) analysis of the surface browning and nutritive value (e.g. available lysine and protein quality) of pizza shells under various processing conditions (Heyne et al., 1985; Unklesbay et al., 1983). These applications are largely limited to simple monochrome tasks.

Recent advances in color machine technology have provided sophisticated tools with extensive new applications for evaluating food quality. Shearer and Payne (1990) reported on sorting bell peppers based on color as characterized by the hue of light reflected from the pepper surface. Shearer and Holmes (1990) proposed a color texture approach to identify seven common cultivars of nursery stock based on co-occurrence matrices. The addition of hue texture improved the classification accuracy from 81.7% to 90.9% over using intensity texture features alone. Liao et al. (1991) used color differences between vitreous and floury endosperm area of a corn kernel to classify corn hardness. Edan et al. (1994) used machine vision extracted appearance attributes such as size, shape and color for a multi-sensor based quality evaluation of tomatoes. These researches used color as an added dimension of information for classification, but made no attempt to quantify the colors of the food materials they were investigating.

There are growing commercial interest in using color machine vision for food sorting. Machine vision based sorting devices are being introduced to food industries. Major products include Merlin Color Sorter (Agri-Tech, Woodstock, VA), GoldRush Color-Sizer (Sunkist Growers, Inc., Ontario, CA), and Tegra (Key Technology, Inc., Walla Wallaawala, WA). Most of the products are not designed for identifying colors as perceived by consumers; however, they are used to classify food products based on color differences.

On the other hand, researchers concerned about customers' response to food colors have studied color change of various green vegetables under different storage conditions. Gunasekaran et al., (1992) compared color measuring results from instrumental and sensory methods of four green vegetables. A tri-stimulus colorimeter was used for their color measurements and data were collected to describe color changes during the storage. To compensate surface pigmentation difference of the vegetables, each data set was the average of 4-8 colorimeter readings taken from different locations on the vegetables. There was no effort in studying the relationship between external color and internal kinetics of quantitative change of pigments.

The color of foods is attributed to the presence of natural pigments (including hemes, myoglobins, chlorophylls, anthocyanins, carotenoids, betacyanins) and added colorants. During food processing and storage, color changes result from the degradation and/or the change to various chemical forms of these pigments. A brown color may also form due to interactions of food components, caused by processing and storage (e.g. nonenzymatic Maillard browning reaction, caramelization, ascorbic acid oxidation). To evaluate the color quality of foods and the kinetics and mechanisms of color change during processing and storage, precision instrumental methods are needed to quantify these pigments and their reaction products. However, most available methods are tedious, time-consuming and expensive, with no possibility of high resolution, on-line control. A sensing technique will be invaluable if pigment contents in food materials can be determined nondestructively and on-line during processing. So far, there is no available information on the relationship between pigments (e.g. chlorophylls, carotenoids, and anthocyanins) contents and machine vision measured colors.

To improve the appearance of color by controlling color development, it is important to establish two important correlations: human perception of color vs. instrumental measurement of colors and instrument measured external colors vs. principal pigment content in the food system. A strong correlation exists between machine vision measured color and human perceived color with trained panel studies (Ling and Tepper, 1995). Knowing that color of food can be described by naturally presented pigments and added colorants, a relationship may be sought between externally measured colors and internal pigment contents for rapid off-line or on-line color evaluation of food materials.

Limited research has been done in the area of using CMV for color measurement and control of food processing and storage. The paramount requirements for developing a useful application of CMV technology lies in the capability to (1) acquire accurate color attributes, (2) quantify pigments and color changes in foods, and (3) establish correlations between pigment contents and direct CMV measurements.

The goal of this research was to develop a rapid color monitoring technology for off-line, on-line processing of value-added food products. Color machine vision technology was used as a fast non-contact, non-destructive method for quantifying and predicting the color quality of processed foods. Specific objectives were to:

(1)    Design an image acquisition system to collect reliable, representative red, green and blue color signals from selected food materials.

(2)    Determine the relationship between the amount of dominant pigments and measured color in selected food materials. Carrot and reformed beef cube were selected as model food systems and the pigments of interest are carotenoids and heme pigments, respectively.

(3)    Evaluate the feasibility of using color for on-line processing control applications.

## METHODOLOGY AND MATERIALS

Accurate machine vision color measurement is a non-trivial task and every step in a color quantification process can affect the accuracy of the measurement. Major areas of the color machine vision system (CMVS) development task consists of acquisition of color images, low-level image processing to compensate for aberrations caused by optical impurity and noisy electronic devices in a system, color quality index selection/development, and defining relationships between instrumental and sensory evaluation results. Successes or failures in developing each of these CMVS building blocks will affect the overall system performance.

*Color Machine Vision System Design*

Color image acquisition model

A better understanding of an image acquisition process is key to accurate color measurement. In essence, instrumental color measurement simulates physical transformations taking place in a human visual system. It is a process of decomposing a flux of radiant energy from a target object into three components, corresponding to the three primary colors, of a sensing device. The decomposition is usually done by optically splitting the flux into three distinct parts and directing each of them toward a corresponding photosensor coupled with a specified optical filter. For the purpose of compatibility, commercial color sensors are made to have tri-stimulus curves similar to standard observers' that are defined by CIE (Commission Internationale de l'Eclairage). CIE has adopted several sets of tri-stimulus curves which comprise bases of standard mathematical color solids such as RGB, XYZ, Lab, and $L^*a^*b^*$.

Regardless of which color solid is used, color signals ($I_i$) generated by an *ideal* machine vision color measurement system at a spatial location $x,y$ can be in general described by an expression given by Horn (1986);

$$I_i(x,y) = \int_{\lambda_1}^{\lambda_2} E(\lambda) * O(\lambda, x, y) * F_i(\lambda) * S(\lambda) \qquad \text{Eq. (1)}$$

where  E($\lambda$) is radiant energy incident upon the object surface,
       O($\lambda,x,y$) - the object spectral reflectance,
       F($\lambda$) - optical filter transmittance,
       S($\lambda$) - sensor spectral sensitivity,
       $\lambda$ - wavelength,
       $i$ - index spanning the color solid stimulus (R,G,B, for example).

Major assumptions in this model are that light source and system spectral parameters are constants and the only variable in a color sensing system is reflective energy from sample objects. Due to less than perfect optical and electronic components in a system, output signals deviate from values predicted by Eq. (1). It is necessary to build a performance based model to define the limitations of color image acquisition with commercially available equipment. A performance model was developed to address important and realistic considerations in building a color image acquisition system (Ruzhitsky and Ling, 1993). Parameters considered in the model include the stability and spectral quality of a light source, sensitivity of sensors, balancing of a sensor's tri-stimulus (RGB) outputs and balancing of R, G and B channels of a color frame grabber in a color machine vision system. The new model is an extension of Eq. (1) and has the form of

$$x, y) = k_i * [\int_{\lambda_1}^{\lambda_2} E(\lambda, x, y, t, v) * O(\lambda, x, y) * L(\lambda, x, y) * F_i \qquad \text{Eq. (2)}$$
$$* S(\lambda, x, y, t) * d\lambda + I_{dc_i}]$$

where $k_i$ - video amplifier gain,
$t$ - time variable,
$v$ - light source power level,
$L$ - lens system
$I_{dci}$ - sensor dark current component.

This model implicitly expresses all major factors affecting image formation in a real CMVS. When the physics behind the image formation is better understood, the information gathered by a CMVS can be better utilized.

Image acquisition system design

Figure 1 is a schematic diagram depicting important components in a color image acquisition system. The halogen light source ($E_\lambda$) in the system produced a spectral quality similar to an illuminant "A" source. A closed-loop controlled lighting system was implemented to provide stable light intensity and spectral quality for the illumination. Feedback photo sensors were used to detect line voltage change or spectral quality degradation of the light source and cause the power regulators to make adjustments to maintain stable and uniform illumination. This system achieved a less than 3 % irradiance deviation in the region of interest.

A CMVS for color measurement was designed around a Sony XC-711 single CCD RGB camera (Sony Corp., Paramus, NJ) and an ITI Series 151 machine vision system (Imaging Technology Inc., Woburn, MA). The color camera had external gain, gamma and balancing adjustments and color temperature selections of either 2800°K or 5600°K. To allow simultaneous capture of three video signals, ADD dual channel digitizer (Scentech, Inc., Woburn, MA) was added to the ITI 151 system. A Fujinon TV Zoom H6x12.5R lens (Fujinon Inc., Wayne, NJ) was attached to the camera. A Minolta Chroma Meter (Minolta Camera Co., Osaka, Japan) which is a general purpose hand-held tri-stimulus colorimeter was used to obtain independent $L^*, a^*, b^*$ values of color samples.

The evaluation of the color measurement using the CMVS was performed using Munsell standard color plates. The tri-stimulus values of the standard color plates were defined for "C" light source; therefore, a "C" light source should be used for the color machine vision measurement. The light source "A" was converted to a light source "C" using color temperature conversion filters (03FCG-259 and 03FCG-261: Melles Griot, CA). The "C" light source has a color temperature close to $6700°K$. A color temperature conversion filter to convert from illuminant "A" to illuminant "C" is shown in Figure 2. Three spectral curves on the right hand side of the figure, from top to bottom, represent the spectral characteristics of (1) the energy received by the camera sensor, (2) the color temperature conversion filter, and (3) an illuminant "A".

The flow chart of the color acquisition procedure is given in Figure 3. The color machine vision system should be white balanced before taking measurement of samples. This can be accomplished by first measuring a white reference target to establish a base line. The R,G,B values of a sample are then acquired. To compare the machine vision measured tri-stimulus values against that of published standard color plates, R,G,B values from CMVS were transformed to $L^*\ a^*\ b^*$ values. This transformation required two steps: first, R,G,B is converted to X,Y,Z by Eq. 3. Next, Eq. 4 is used to convert X,Y,Z to $L^*, a^*, b^*$ color coordinates.

Two step RGB - L a b transformation

$$\begin{bmatrix} X \\ Y \\ Z \end{bmatrix} = \begin{bmatrix} 0.490 & 0.310 & 0.200 \\ 0.177 & 0.813 & 0.011 \\ 0.000 & 0.010 & 0.990 \end{bmatrix} \begin{bmatrix} R \\ G \\ B \end{bmatrix}$$

**Eq. (3)**

for $1 \le 100\ Y \le 100$

$$L^* = 25\ (\frac{100\ Y}{Y_o})^{\frac{1}{3}} - 16, \quad a^* = 500 \left[ \left( \frac{X}{X_o} \right)^{1/3} \right.$$

$X_o, Y_o, Z_o = tristimulus\ value$

**Eq. (4)**

Color image acquisition tuning

White balancing is another essential step in achieving accurate color measurement and it can be done by adjusting the sensor's output and the digitizer's

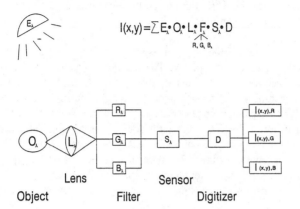

$$I(x,y) = \sum_\lambda E_\lambda \cdot O_\lambda \cdot L_\lambda \cdot F_\lambda \cdot S_\lambda \cdot D$$

Figure 1. A schematic diagram of the color image acquisition system. An image acquisition system consists of many optical and electronic components. A good understanding of components' performances in a system is critical to the assurance of accurate color measurement.

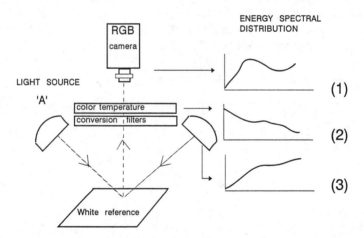

Figure 2. A schematic drawing of the color temperature conversion process. The spectral distribution of the light source, which was a standard 'A' light source, was converted to 'C' light source using a set of two color temperature conversion filters. The color temperature conversion was necessary for accurate color measurement. Figures 2-(1), 2-(2), and 2-(3) are the spectral characteristics of light source 'A', color conversion filters, and converted light source 'C', respectively.

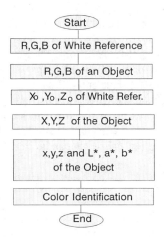

Figure 3. A flow chart of the color acquisition procedure. The RGB signals of a white reference and samples were first acquired by a camera sensor. Colors of samples were then acquired and converted from RGB color coordinate to $L^*a^*b^*$ color space for color identification.

output. A number of camera parameters, such as auto-gain, gamma, and color temperature, contribute to the accuracy of the measurement. The auto-gain should be disabled and gamma set to 1. While viewing a white reference material, camera was adjusted so that each video signal, R,G,B, had same level. These signals were monitored with an oscilloscope. The outputs from the R,G,B frame grabbers had the same gray levels by adjusting the gains and offsets for each video channel.

Linearization of a color machine vision system is another important consideration. It is especially important when ratios between two color channels or linearcombinations of information from multiple channels are to be used to characterize spectral features in a scene. It is desirable for a linear relationship between CMVS outputs (gray level values) and energy received by the sensor. This was achieved by measuring dark currents and determining the relationship between the input energy level and the output gray level of each of the three channels. Dark currents were measured by closing the camera aperture and reading gray level values from each channel. Data collection for the latter was achieved by changing camera lens aperture settings while the scene irradiance was maintained constant during the experiment. Transfer functions of the R,G,B channels of the CMVS were determined through statistical regression analyses. A detailed description of the linearization process is given by Ruzhitsky and Ling (1992).

### Evaluation of color measurement accuracy

The color image acquisition system was evaluated in two ways. First, finding the correlation between CMVS and colorimeter readings of the color standards.

Second, sensitivity of the system was defined to determine how well the CMVS can resolve color differences. A set of color standards, produced by Macbeth Corp. and specified in Munsell notations (Macbeth Corp., Newburgh, NY), was used for the evaluation. The set consists of seven color groups: 7.7R 4.57/16.9 (red), 9.0YR 7.42/13.6 (buff), 3.8BG 4.73/1.1 (blue gray), 0.2Y 8.28/3.8 (beige), Sunset orange, 3.1YR 6.45/14.5 (orange), 5.6GY 6.25/0.33 (smoke gray). All but the 9.0YR group, which has six, consist of seven colors of minor difference. The color differences among two extreme colors within groups range from 1.2 to 5.5 CIE units. A CIE unit of color difference is one unit of $\Delta E$ which can be calculated from Eq. (5).

$$\Delta E = \sqrt{\Delta L^{*2} + \Delta a^{*2} + \Delta b^{*2}}$$

$$where \ \Delta L^* = L^*_{color \ i} - L^*_{color \ j}$$

$$\Delta a^* = a^*_{color \ i} - a^*_{color \ j} \qquad \qquad Eq. \ (5)$$

$$\Delta b^* = b^*_{color \ i} - b^*_{color \ j}$$

whereas *color i*, *color j* are any two colors of interest.

The CMVS readings were taken at the center of the image to avoid influences from optical axis offset and optical aberrations. Sensing instability of the system was accounted for by (1) taking multiple readings from the same sample location over time and (2) averaging color reading of the sampling window which was 25x25 pixel. Averaged R,G,B readings from the CMVS were transformed to $L^*,a^*,b^*$ color coordinates to compare against the colorimeter readings.

CMVS measurements were correlated with a Minolta Chroma Meter for color plates measurements. Tri-stimulus values of six Munsell color standard plates were obtained by the color machine vision system and the Minolta Chroma Meter. Table 1 gives the tri-stimulus values of Munsell color standard plates obtained from two color measuring approaches. Notice that the values are generally in agreement which indicates that (1) the color machine vision system is capable of measuring standard color plates, and (2) good agreements were found between the Minolta Chroma Meter readings and the designated tri-stimulus values of the standard color plates.

*Pigment Quantification*

Sample preparation

The food systems studied were salty carrots and salty beef cubes processed in the CRAMTD Food Manufacturing Facility at Rutgers University (Piscataway, NJ) and carrot puree. Salty carrots were prepared by placing 92 g carrot cubes per pouch with 135 g brine solution per pouch (0.86% final salt concentration). Salty beef cubes were prepared by placing 91 g diced reformed beef cubes per pouch (beef obtained by Huttenbauer, Cincinnati, OH) with 133 gram brine per pouch. Approximately 150 pouches were sealed and processed at 8 different time/temperature combinations. Carrot puree was prepared by first cooking frozen, diced carrot cubes in boiling water. The samples were boiled from 30 to 180 minutes with a 30 minutes interval. Purees were then prepared from the boiled carrots. A total of six carrot puree samples were prepared.

## Extraction and determination of carotenoids

The procedure used for the extraction of $\alpha$-carotene and ß-carotene from carrots was based on the Bureau and Bushway method (1986). Fresh carrots were homogenized in a Waring blender and divided into 4 parts; each part was processed at 121°C in a Presto pressure cooker for different durations. After treatment, samples were vacuum-dried for 48 hrs at $T < 20°C$. One gram dry sample was extracted with 20 ml tetrahydrofurane (THF) containing 0.1% ascorbic acid and the extract was vacuum-filtered through Whatman No.42 filter paper. The filter cake was re-extracted using 25 ml THF to remove all carotenoids. All extracts were combined into a flask and filled up to 50 ml volume with THF. The same extraction procedure was used for salty carrots.

The determination of $\alpha$-carotene, trans- and cis-ß-carotene from salty carrots was done by reverse-phase HPLC. All trans-, $\alpha$- and ß-carotene used for the generation of calibration curves were purchased from Sigma Chemical (St.Louis, MO). A Waters Nova-Pak C-18 column (pore size 4 $\mu$m; dimensions 3.9 mm * 300 mm) with mobile phase acetonitrile:methanol:THF: ammonium acetate (in methanol) 35:56:7:2 and flow rate 2 ml/min was used in the HPLC. Twenty-five $\mu$l extract was injected into the column. The mobile phase was more effective in separating $\alpha$-carotene from trans- and cis-ß-carotene from the mixture of carotenoids compared to the mobile phase of acetonitrile:ethyl acetate:methanol:ethyl acetate (Craft, 1992). Complete separation of trans-$\alpha$-carotene from ß-carotene was achieved; however, only partial separation of trans- and cis-ß-carotene was possible.

## Heme extraction and determination from beef

A modified classical method (Hornsey, 1956) was used for extracting heme. Brine was removed by coarse filtering. Beef cubes were minced in a Waring blender with minimum delay in a dark room using dry ice during mincing. This procedure improved the extraction by an acetone-based solvent (80:4:2 acetone:water: HCl). After mixing to a smooth paste and placing the tubes on a shaker for 1 hr, 10 grams of minced meat were extracted with 43 ml of the solvent. The amounts used were based on 70% meat moisture and resulted in 80% (v/v) acetone in the final solution. After filtering, we recorded the visible spectra of the filtrates; we also monitored the absorbance peaks at 510, 540 and 640 nm. The heme concentration was calculated after the generation of a calibration curve at 640 nm using hemin chloride as a standard (obtained from Sigma Chemical, St.Louis, MO).

RESULTS

*Color Machine Vision System Performance*

The CMVS's color resolution is color dependent. At a 95% confidence interval, the CMVS was able to resolve 0.7 CIE unit color difference in the 3.8BG group. On the other hand, at the same confidence interval, the CMVS can only resolve 3.1 units of color differences in group 9.0YR. It was not possible to determine the color measurement resolution in the 7.7R group due to a larger measurement uncertainty and a smaller color difference, 1.5 CIE unit measured by the colorimeter, between two extreme colors in the group. The color measurement resolution of the CMVS was mostly affected by the stability of the color measurements. Figure 4 shows the uncertainty involved in color measurements. One hundred readings were obtained in sequence from the same sample under the same image acquisition setup. Each point represents the standard deviation of 100 readings from 1 color plate. The standard deviations varies from 0.8 to 2.0 among various color groups. The smaller the deviation, the better a CMVS can resolve colors.

There is an excellent correlation between the color differences of standard color chips measured by the CMVS and the colorimeter. Using white color standard as a reference, the color differences were calculated between each color measurement and the reference (Figure 5). A strong correlation exists between the measurements (R > 0.98). Although groups 9.0YR-buff and 0.2Y-beige are relatively close to each

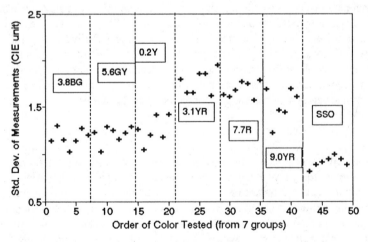

Figure 4. The color measurement stability of the color machine vision system. The CMVS has a range of color resolution in measuring various color due to various level of noise in measuring individual colors. The graph shows the magnitude of measurement deviation of each color from 100 repeated measurements.

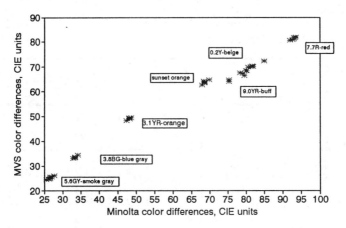

Figure 5. The color machine vision system and the colorimeter have an R value better than 0.98, in determining color differences between sample colors and a white standard.

other in this graph, they have a color difference of 55.08 CIE units between each other. Two colors can have similar color differences relative to the reference while being located at the opposite sides of the white in a color coordinate system. Figure 6 shows distribution of the color plates in the $L^*a^*b^*$ color coordinate system.

Correlation was established between the colors of food samples measured by the colorimeter color and the color machine vision system. The correlation coefficients between $L^*a^*b^*$ color parameters for beef and carrots obtained from the two instruments can be seen in Table 2. For all the data collected, a high correlation was observed ($R >= 0.98$). The established correlation makes it possible to verify CMVS's color measurement accuracy of colors that are not available from published color standards. Furthermore, the correlations between the CMVS and colorants in food systems can also be inferenced based on established correlations between colorimeter measurements and colorant contents.

*Color Change Kinetics*

Reformed beef cube study

Changes in color of salty beef cubes during processing at three processing temperatures are shown in Table 3. Only minor color differences were observed among beef samples processed at the same temperature but for different times. The range for color changes is small in this study, since significant color changes have already occurred during the production of the reformed beef. A t-test (statistical test for significant difference) for the $L^*a^*b^*$ data series revealed significant differences ($p < 0.01$) only for the $a^*$ value.

Table 1. The tri-stimulus values of Munsell color standard plates obtained from two color measuring approaches.

| Color Standard | Minolta | | | Macbeth Data | | | MVS source 'C' | | |
|---|---|---|---|---|---|---|---|---|---|
| | x | y | z | x | y | z | x | y | z |
| 5GY 6/4 | 0.342 | 0.398 | 0.26 | 0.346 | 0.402 | 0.252 | 0.353 | 0.322 | 0.325 |
| 10R 6/12 | 0.51 | 0.356 | 0.134 | 0.512 | 0.367 | 0.121 | 0.408 | 0.286 | 0.306 |
| 5YR 7/12 | 0.498 | 0.397 | 0.105 | 0.501 | 0.408 | 0.091 | 0.42 | 0.322 | 0.258 |
| 7.5YR 4/4 | 0.408 | 0.369 | 0.223 | 0.425 | 0.383 | 0.192 | 0.369 | 0.269 | 0.362 |
| 7.5YR 6/4 | 0.386 | 0.361 | 0.253 | 0.388 | 0.366 | 0.246 | 0.367 | 0.305 | 0.328 |
| 7.5YR 8/4 | 0.373 | 0.359 | 0.268 | 0.372 | 0.36 | 0.268 | 0.362 | 0.33 | 0.308 |

Table 2. Correlation coefficients between $L^*a^*b^*$ color parameters for beef and carrots measured by Minolta Chroma Meter and the CMVS.

| Processed samples | L* | | a* | | b* | |
|---|---|---|---|---|---|---|
| | Minolta | CMVS | Minolta | CMVS | Minolta | CMVS |
| Carrots, 13min@245°F | 49.70 | 64.58 | 21.54 | 46.99 | 44.69 | 0.88 |
| Carrots, 17min@250°F | 50.34 | 64.24 | 23.53 | 49.84 | 46.25 | -0.93 |
| Carrots, 21min@255°F | 49.79 | 62.95 | 20.29 | 48.60 | 46.41 | -1.77 |
| Beef, 13min@245°F | 44.37 | 55.62 | 10.25 | 37.30 | 12.39 | -18.36 |
| Beef, 17min@250°F | 44.47 | 54.67 | 10.79 | 37.68 | 12.61 | -18.37 |
| Beef, 21min@255°F | 44.08 | 54.81 | 11.25 | 36.70 | 13.53 | -17.31 |
| Correlation Coefficient | 0.9901 | | 0.9846/ | | 0.9923 | |

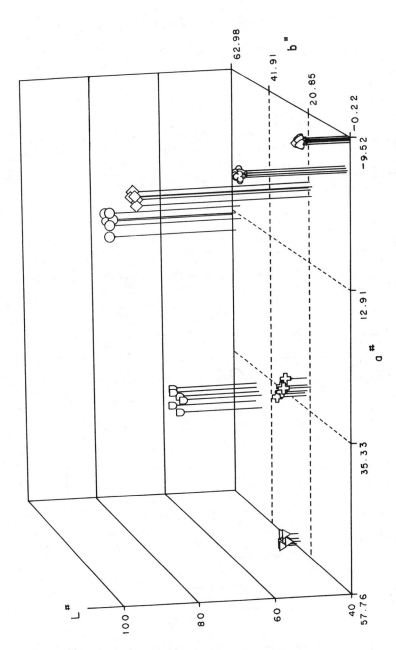

Figure 6. A distribution of the color machine vision system measured colors in $L^*a^*b^*$ color system. There are seven major groups with major color differences. Each group consists of six or seven colors with minor color differences.

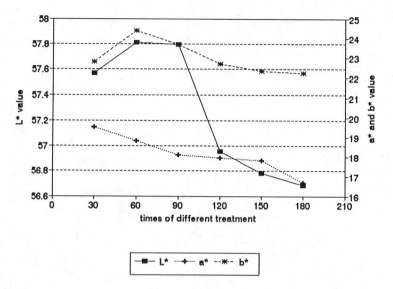

Figure 7. Color change of carrot puree due to the thermal processing. Color attribute L*, a*, and b* are presented in the graph.

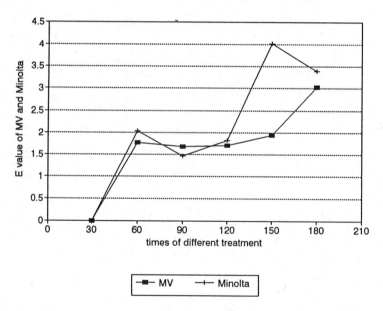

Figure 8. Excellent agreement was found between the color machine vision and the colorimeter measured carrot puree colors. The largest disagreement was less than 2 CIE units, while most differences are less than 0.5 CIE unit.

Variability is not the same for all the parameters studied. Using standard deviation (STD) as a measure of variability, $L^*$ value was found the most variable parameter (STD 1.74-3.22), whereas $a^*$ value was the least variable (STD 0.46-0.79). The changes of $a^*$ ($\Delta a^*$) exhibited a strong positive correlation with processing time for all temperatures studied (all R values are greater than 0.99). On the other hand, $L^*$ exhibited a decreasing trend (R -0.86 to -0.89), whereas $b^*$ was not significantly correlated with processing time. $\Delta E$, which is a measure of the sum of the color changes expressed as CIE units, was highly correlated with processing time in all cases (all R values are greater than 0.97), mainly as a result of $a^*$ change. Finally, it is noteworthy that the correlation coefficients did not vary significantly with processing temperature. Furthermore, the normality of the color data was verified using Shapiro-Wilc W test. It is found that $a^*$ and $b^*$ values consistently followed normal distribution, (9/9 and 8/9 of the cases examined respectively) whereas the distribution of $L^*$ was not considered normal for almost half of the cases examined (5/9).

From all the observations above, it is clear that the $a^*$ parameter, which serves as an index of redness/greenness is the least variable color parameter. It follows a normal distribution and can be used to distinguish between different processing times due to its high correlation coefficient with processing time. Therefore, accurate monitoring of the $a^*$ value can provide useful and meaningful information about the quality of processed beef color and the extent of processing. Finally, the $\Delta E$ value may prove to be very helpful as it generally correlates very well with processing time in all cases and encompasses the changes in all color parameters.

### Carrot puree study

Color change of carrot puree due to the thermal processing was also quantified. The best color attribute that may be used as a feedback for the thermal process control was $a^*$. As shown in Figure 7, $a^*$ values consistently decreased value as processing time increased. $L^*$ and $b^*$, on the other hand, slightly increased between 30 and 60 minutes of processing and then steadily decreased with the processing time. Overall change of each of the components was within 3 CIE units.

The experiment also confirmed the high correlation between the colorimeter (Minolta Chroma Meter) and CMVS measured colors of food materials. Figure 8 shows small differences between the colorimeter and the CMVS measured color differences due to the thermal processing. The measurements were taken from pureed, diced carrot cooked for various time durations. One can see that the largest disagreement between the colorimeter and the CMV measured carrot puree color difference was about 2 CIE unit while most disagreements are within 0.5 CIE unit. The larger difference could be an outlier and should be further investigated.

The results from the study of processed salty carrots suggested that the external color of the salty carrot samples is related to the extent of thermal processing (Table 4). The results also indicate that $\Delta a^*$ is the parameter that correlates stronger

Table 3. Correlation of colo parameter differences with processing time for beef samples processed at various time-temperature combination.

| Processing T (°F) | Processing time (min) | L* | a* | b* | ΔL* | Δa* | Δb* | ΔE |
|---|---|---|---|---|---|---|---|---|
| No Processing | 0 | 45.79±2.55 | 9.03±0.65 | 13.07±0.79 | 0.00 | 0.00 | 0.00 | 0.00 |
| 245 | 13 | 44.37±1.94 | 10.55±0.54 | 12.21±0.96 | -1.42 | 1.55 | -0.86 | 2.27 |
| | 17 | 44.71±2.74 | 11.03±0.76 | 14.07±1.26 | -1.08 | 2.00 | 1.00 | 2.48 |
| | Correlation coefficient | | | | -0.897 | 0.999 | 0.994 | 0.991 |
| | 13 | 45.53±2.13 | 10.49±0.79 | 14.37±1.03 | -0.26 | 1.46 | 1.30 | 1.98 |
| 250 | 17 | 44.47±2.36 | 10.78±0.60 | 12.61±1.00 | -1.32 | 1.76 | -0.46 | 2.24 |
| | 21 | 44.60±1.82 | 11.07±0.70 | 13.73±1.03 | -1.19 | 2.04 | 0.66 | 2.45 |
| | Correlation coefficient | | | | -0.860 | 0.994 | 0.173 | 0.981 |
| | 13 | 44.91±3.22 | 10.60±0.76 | 14.54±1.68 | -.088 | 1.57 | 1.47 | 2.32 |
| 255 | 17 | 43.49±2.59 | 10.73±0.63 | 12.92±0.90 | -2.30 | 1.70 | -0.15 | 2.86 |
| | 21 | 44.08±1.74 | 11.24±0.46 | 13.53±0.86 | -1.71 | 2.21 | 0.46 | 2.83 |
| | Correlation coefficient | | | | -0.881 | 0.991 | 0.173 | 0.973 |

Table 4. Correlation of color parameter differences with processing time for carrot samples processed at 245°F.

| Processing time (min) | L* | a* | b* | ΔL* | Δa* | Δb* | ΔE |
|---|---|---|---|---|---|---|---|
| 0 | 52.40±0.97 | 27.99±1.02 | 45.39±1.20 | 0.00 | 0.00 | 0.00 | 0.00 |
| 13 | 49.70±0.59 | 21.54±0.75 | 44.69±1.18 | -2.70 | -6.45 | -0.70 | 7.03 |
| 17 | 50.34±0.50 | 23.53±0.48 | 46.25±0.76 | -2.06 | -4.46 | 0.86 | 4.99 |
| 21 | 49.79±0.79 | 20.29±0.67 | 46.41±0.85 | -2.61 | -7.70 | 1.02 | 8.19 |
| Correlation coefficient | | | | -0.899 | -0.912 | 0.546 | 0.841 |

with processing time, showing a decreasing trend. This decrease in the $a^*$ value may be related to heat degradation of carotenoids, the main pigments of carrots that are responsible for the orange color. A somewhat linear relationship exists between $\Delta b^*$ values and the duration of the thermal process. Furthermore, the color changes due to heating in carrots are more extensive than in beef, as judged after comparing the $\Delta E$ value for the two food systems ($\Delta E$ is 1.98-2.86 for beef and 4.99-8.19 for carrots).

### Pigment Degradation Studies

The pigment chemistry of the meat systems suggests that the changes in the $a^*$ parameter are correlated to changes that occur in the concentrations of myoglobin and heme as these are the main pigments in meat and are responsible for the red color of fresh meat. Heat causes degradation of the heme ring with subsequent release of the iron moiety in meat (Shricker and Miller, 1982). This phenomenon, apart from nutritional importance, significantly changes meat color. To find a correlation between externally measured color and internal pigments or pigment related compounds, a chemical basis for the color changes due to processing must be established. The principal chemical markers selected should be the major source of color for the systems studied. Heme was selected as a marker for the beef and $\alpha$- and ß-carotene for the carrots.

### Quantitative determination of heme degradation

We performed heme determinations on the beef samples to follow the heat degradation of the pigment related compound. The results for the samples processed at 250 and 255°F are shown in Figure 9. The results indicate that significant heme degradation occurs with processing; this degradation relates with processing time for the temperatures employed and thus it may be a suitable processing marker. Although the scarcity of time/temperature combinations does not allow for conclusions on the kinetics of the degradation, zero and first-order models fit well to the available data. Assuming a zero order degradation, R is 0.95 and 0.98 for 250 and 255°F respectively; approximately the same numbers were obtained for a first-order model. These results are encouraging considering the uneven heat distribution in the retort that created "hot" and "cold" spots in the cooker. In addition, the thermal profile is far less than ideal due to long warm-up time, which is required to reach preset processing temperature, and cooling time.

The color parameters $a^*$ and $\Delta a^*$ were highly correlated with the concentration of heme in the beef samples (Figure 10). A high correlation coefficient of 0.97 indicated that the $a^*$ change during processing is strongly correlated to the degradation of heme, the chromophore ring of myoglobin which represents the most significant pigment in meat.

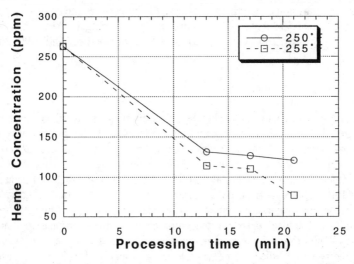

Figure 9. Heme degradation during processing of beef samples at 250 and 255°F.

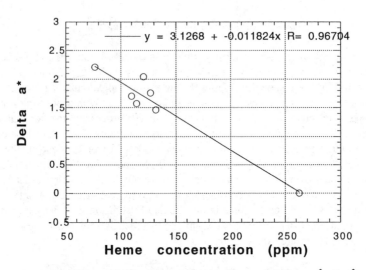

Figure 10. Correlation of the change of color parameter a* ($\Delta$a*) during processing with the concentration of heme in beef samples.

Quantitative determination of carotenoids' degradation

The results indicate that processing time affected the concentration of α-carotene and ß-carotene due to heat degradation (Table 5). Note that the peak area of ß-carotene represents the total peak area of trans- and cis-ß-carotene.

The change in concentration of carotenes against time at a temperature of 255°F is shown in Figure 11. The degradation pattern is very similar for both compounds. We observed an excellent correlation of the degradation of carotenoids with time, with R ~ 0.99 for both compounds, an indication of a zero-order reaction. In contrast, the first order model does not fit as well (R ~ 0.96). We anticipate that the carotenoid degradation is the chemical basis for the decrease of the a* value during processing as detected by the colorimeter and CMVS measurements. Finally, our results suggest that the heat degradation patterns of α- and ß-carotene renders themselves as suitable compounds to serve as kinetic markers for the processing monitoring.

## CONCLUSIONS

In-line/on-line feedback control of color of food during processing can improve not only color quality but also color related quality such as texture and appearance. To do this, there are three major aspects: development of an in-line/on-line color sensor; understanding of color change kinetics; and establish correlations between instrumental measured and sensory panel perceived colors of foods. In this research, we have chosen color machine vision technology for the measurement of colors of food due to its superior spatial resolution over conventional instruments such as colorimeter or spectrophotometer. Relationships between measured colors and corresponding principal chemical markers were established for the model food systems. We have also found excellent correlations between the color machine vision system (CMVS) measured and a sensory panel determined colors of food samples (Ling and Tepper, 1995). We believed that a CMVS can be used for food process control to ensure color quality as perceived by consumers.

Color machine vision systems are readily available from commercial vendors. The systems, however, are mostly designed for graphical presentation or color classification tasks and are not suitable for color measurement or color quantification tasks. In considering a CMVS for color measurement, users must know the factors that affect color measurement including color temperature of the light source, balancing of color signals, and proper selection of color coordinates to represent colors. Furthermore, proper selection and usage of a color camera sensor is critical. To obtain linear, reliable, and accurate outputs from a color camera, camera settings such as auto-gain, gamma, and color temperature should be properly selected. Color image acquisition is a complex process and one should be aware of critical factors affecting color measurement tasks. Therefore, a good understanding and control of a color image acquisition process is essential in assuring accurate color measurements.

Table 5. α- and ß- carotene concentrations in raw and heated carrots.

| Proc.time at 255°F (min) | α-carotene | | ß-carotene | |
|---|---|---|---|---|
| | Concentration (μg/g dry sample) | Retention (%) | Concentration (μg/g dry sample) | Retention (%) |
| 0 | 849.78 | 100.0 | 2841.00 | 100.0 |
| 13 | 430.77 | 50.7 | 1579.19 | 55.6 |
| 17 | 225.66 | 26.6 | 854.57 | 30.1 |
| 21 | 169.52 | 19.9 | 637.97 | 22.5 |

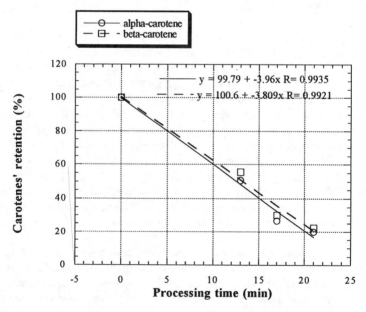

Figure 11. The α- and ß-carotene carotenes concentrations degradation of salty carrots due to the thermal process as functions of time at a temperature of 255°F.

A CMVS was designed and evaluated on colors perceived by human. The machine vision measured tri-stimulus values in R,G,B color coordinate were transformed to L*a*b* color space for better representation of human perceived colors. The CMVS measurements were evaluated using Munsell standard color chips, a colorimeter, and a sensory panel. Excellent correlations were found between the CMVS and the colorimeter, which is designed to quantify human perceived colors. The CMVS's ability in quantifying human perceived colors was evidenced by accurate measurement of Munsell standard color chips and the excellent correlations with the sensory panel perceived colors (R > 0.90). From the experiments, it was possible to resolve up to 0.7 CIE unit color difference of the tested colors.

The established correlation between the CMVS and the colorimeter for homogeneous color measurement demonstrated the potential of using CMV for color kinetics studies. Correlations between CMV measured color and concentration of pigments or pigment related compounds in food systems can be determined based on the established relationships using colorimeter. Nonetheless, CMV will be able to provide more detailed information due to its high spatial resolution. Instead of just one averaged reading of samples provided by colorimeter, CMV provides color information of more than 80,000 locations on samples. CMV can be seen as an upward compatible color sensing technology for color measurement of food materials. CMV technology should be further developed to quantify other attributes of appearance quality such as distributions of color, shape, and size of ingredients in food systems.

Our research results also have demonstrated the feasibility of using CMV as a fast, non-destructive, in-line/on-line color sensor for a closed-loop controlled food process to produce high color quality food products. CMV was used, indirectly, to relate sensory panel perceived color quality and major pigment contents in food systems during thermal processes. The correlations between CMVS measured colors and human perceived colors of selected food models were established; chemical markers were identified for color changes in thermal processes; the correlations between external colors and principal pigment contents were established for two model food systems: carrots and reformed beef cube.

This study also identified and verified the chemical basis of color change during retort thermal processing. Concentrations of certain food pigments correlates well with external color parameters and especially parameter a* (index of redness-greenness). This observation, along with the dependence of pigment concentration versus processing time provides an excellent opportunity for process control using CMVS. Finally, identification of the compounds responsible for color change provided an opportunity to manipulate chemical and/or physical properties of the samples in order to control the extent and kinetics of color change that will directly influence color quality of the food during processing.

This approach can also be used in other food systems for which color is an important quality parameter. For example, food containing Maillard reaction products

CHEMICAL MARKERS FOR PROCESSED AND STORED FOODS

(cookies, bread, potatoes, cereal), anthocyanins (grapes, apple and cranberry), chlorophyll (peas, spinach, broccoli), iron compounds (iron fortified rice, cereal) are good candidates for future studies. In addition to the proven technical feasibility in using the CMVS for color quality control of food materials, we feel that color machine vision technology is becoming an appropriate technology. With dropping computer hardware prices, a personal computer based sensing and control system for color of foods is becoming affordable for mid and small-size food processing plants. The CMV technology will be efficient and economical for quality quantification of colors of foods.


**Acknowledgment**

REFERENCES

Alahakoon P.M.K.; Sudduth, K.A.; Hsieh, F. Image analysis of the internal structure of corn extrudate. **1991**, ASAE paper 916540, St. Joseph, Michigan.

Barrett, A.M.; Peleg, M. Cell size distributions of puffed corn extrudates. *Journal of Food Science*, **1992**, 57(1), 146-148.

Bourne, M.C. Food texture and viscosity. Concept and measurement; Academic Press: New York. **1982**.

Bureau, J.L.; Bushway R.J. HPLC determination of carotenoids in fruits and vegetables in the United States. *Journal of Food Science*, **1986**, 51(1), 128-130.

Chen, P.; Sun, Z. A review of non-destructive methods for quality evaluation and sorting of agricultural products. Journal of Agricultural Engineering Research. **1991**, 49, 85-98.

Clydesdale, F.M. Color as a factor in food choice. *Critical Review in Food Science and Nutrition*, **1993**, 33(1), 83-101.

Craft, N.E. Carotenoid reverse-phase high-performance liquid chromatography methods: reference compendium. *Methods in Enzymology*, **1992**, V. 213, 185-205.

Edan, Y., H.; Pasternak, D. G; Ozer N.; Shmulevich, I.; Rachmani D.; Fallik E.; Grinberg, S. Multi-sensor quality classification of tomatoes. **1994**, ASAE paper 946032. St. Joseph, MI.

Gunasekaran, V.; Shewfelt, R.L.; Chinnan, M.S. Detection of color changes in green vegetables. *Journal of food science*, **1992**, 57(1), 149-154.

Heyne, L.; Unklesbay, N.; Unklesbay, K.; Keller, J. Computerized image analysis of protein quality of simulated pizza crusts. *Canadian Institute of Food Science and Technology Journal*, **1985**, 18(2), 168-173.

Horn, B. K. P. Robot Vision; The MIT Press: Cambridge MA. **1986**.

Hornsey, H.C. The Colour of Cooked Cured Pork. J. Sci. Food Agric., **1956**, 7, 534-540.

IFT (Institute of Food Technologists). Quality of fruits and vegetables- a scientific status summary by Institute of Food Technologists' expert panel on food safety and Nutrition. *Food Technology*, **1990**, 44, 99.

Liao, K.; Reid, J.F.; Paulsen, M.R.; Shaw, E.E. Corn kernel hardness classification by color segmentation. **1991**, ASAE paper 91354. St. Joseph, MI.

Ling, P.P.; Tepper, B.J. Unpublished data. **1995**.

Ni, H.; Gunasekaran, S. Computer vision method for cheese shred morphology evaluation. **1994**, ASAE paper 943504. St. Joseph, MI.

Roudot, A.C. Image analysis of kiwi fruit slices. *Journal of Food Engineering*, **1989**, 9(2), 97-118.

Ruzhitsky, V.N.; Ling, P.P. Ling. Machine vision for tomato seedling inspection. **1992**, ASAE technical No. 927020, St Joseph, MI.

Ruzhitsky, V.N.; Ling, P.P. Development of a Machine Vision Calibration Procedure for Color Measurement. **1993**. ASAE/NABEC paper No.93-206. St. Joseph, MI.

Schricker, B.R.; Miller, D.D. Effect of cooking and chemical treatment on heme and nonheme iron in meat. *Journal of Food Science*, **1982**, 48, 1340-3, 1349.

Shearer, S.A.; Payne, F.A. Color and defect sorting of bell peppers using machine vision. *Transactions of the ASAE*, **1990**, 33(6), 2045-2050.

Shearer, S.A.; Holmes, R.G. Plant identification using color co-occurance matrices. *Transactions of the ASAE*, **1990**, 33(6), 2037-2044.

Smolarz, A.; Van Hecke, E.; Bouvier, J.M. Computerized image analysis and texture of extruded biscuits. *Journal of Texture Studies*, **1989**, 20, 223-234.

Tillett, R.D. Image analysis for agricultural processes: a review of potential opportunities. *Journal of Agricultural Engineering Research*, **1991**, 50, 247-258.

Unklesbay, K.; Unklesbay, N.; Keller, J. Determination of internal color beef ribeye steaks using digital image analysis. *Food Microstructure*, **1986**, 5, 227-231.

Unklesbay, K.; Unklesbay, N.; Keller, J.; Grandcolas, J. Computerized image analysis of surface browning of pizza shell. *Journal of Food Science*, **1983**, 48, 1119-1123.

Zayas, I.; Converse, H.; Steele, J. Discrimination of whole from broken corn kernels with image analysis. *Transactions of the ASAE*, **1990**, 33:(5), 1642-1646.

Zhang, H.; Tan, J.; Gao. X. SEM image processing for food structure analysis. **1994**. ASAE paper 943506. St. Joseph, MI.

# Author Index

# Affiliation Index

# Subject Index

# Highlights from ACS Books

*Good Laboratory Practice Standards: Applications for Field and Laboratory Studies*
Edited by Willa Y. Garner, Maureen S. Barge, and James P. Ussary
ACS Professional Reference Book; 572 pp; clothbound ISBN 0–8412–2192–8

*Silent Spring Revisited*
Edited by Gino J. Marco, Robert M. Hollingworth, and William Durham
214 pp; clothbound ISBN 0–8412–0980–4; paperback ISBN 0–8412–0981–2

*The Microkinetics of Heterogeneous Catalysis*
By James A. Dumesic, Dale F. Rudd, Luis M. Aparicio, James E. Rekoske,
and Andrés A. Treviño
ACS Professional Reference Book; 316 pp; clothbound ISBN 0–8412–2214–2

*Helping Your Child Learn Science*
By Nancy Paulu with Margery Martin; Illustrated by Margaret Scott
58 pp; paperback ISBN 0–8412–2626–1

*Handbook of Chemical Property Estimation Methods*
By Warren J. Lyman, William F. Reehl, and David H. Rosenblatt
960 pp; clothbound ISBN 0–8412–1761–0

*Understanding Chemical Patents: A Guide for the Inventor*
By John T. Maynard and Howard M. Peters
184 pp; clothbound ISBN 0–8412–1997–4; paperback ISBN 0–8412–1998–2

*Spectroscopy of Polymers*
By Jack L. Koenig
ACS Professional Reference Book; 328 pp;
clothbound ISBN 0–8412–1904–4; paperback ISBN 0–8412–1924–9

*Harnessing Biotechnology for the 21st Century*
Edited by Michael R. Ladisch and Arindam Bose
Conference Proceedings Series; 612 pp;
clothbound ISBN 0–8412–2477–3

*From Caveman to Chemist: Circumstances and Achievements*
By Hugh W. Salzberg
300 pp; clothbound ISBN 0–8412–1786–6; paperback ISBN 0–8412–1787–4

*The Green Flame: Surviving Government Secrecy*
By Andrew Dequasie
300 pp; clothbound ISBN 0–8412–1857–9

---

For further information and a free catalog of ACS books, contact:
American Chemical Society
Customer Service & Sales
1155 16th Street, NW, Washington, DC 20036
Telephone 800–227–5558

# Bestsellers from ACS Books

*The ACS Style Guide: A Manual for Authors and Editors*
Edited by Janet S. Dodd
264 pp; clothbound ISBN 0–8412–0917–0; paperback ISBN 0–8412–0943–X

*Understanding Chemical Patents: A Guide for the Inventor*
By John T. Maynard and Howard M. Peters
184 pp; clothbound ISBN 0–8412–1997–4; paperback ISBN 0–8412–1998–2

*Chemical Activities* (student and teacher editions)
By Christie L. Borgford and Lee R. Summerlin
330 pp; spiralbound ISBN 0–8412–1417–4; teacher ed. ISBN 0–8412–1416–6

*Chemical Demonstrations: A Sourcebook for Teachers,*
*Volumes 1 and 2,* Second Edition
Volume 1 by Lee R. Summerlin and James L. Ealy, Jr.;
Vol. 1, 198 pp; spiralbound ISBN 0–8412–1481–6;
Volume 2 by Lee R. Summerlin, Christie L. Borgford, and Julie B. Ealy
Vol. 2, 234 pp; spiralbound ISBN 0–8412–1535–9

*Chemistry and Crime: From Sherlock Holmes to Today's Courtroom*
Edited by Samuel M. Gerber
135 pp; clothbound ISBN 0–8412–0784–4; paperback ISBN 0–8412–0785–2

*Writing the Laboratory Notebook*
By Howard M. Kanare
145 pp; clothbound ISBN 0–8412–0906–5; paperback ISBN 0–8412–0933–2

*Developing a Chemical Hygiene Plan*
By Jay A. Young, Warren K. Kingsley, and George H. Wahl, Jr.
paperback ISBN 0–8412–1876–5

*Introduction to Microwave Sample Preparation: Theory and Practice*
Edited by H. M. Kingston and Lois B. Jassie
263 pp; clothbound ISBN 0–8412–1450–6

*Principles of Environmental Sampling*
Edited by Lawrence H. Keith
ACS Professional Reference Book; 458 pp;
clothbound ISBN 0–8412–1173–6; paperback ISBN 0–8412–1437–9

*Biotechnology and Materials Science: Chemistry for the Future*
Edited by Mary L. Good (Jacqueline K. Barton, Associate Editor)
135 pp; clothbound ISBN 0–8412–1472–7; paperback ISBN 0–8412–1473–5

---

For further information and a free catalog of ACS books, contact:
American Chemical Society
Customer Service & Sales
1155 16th Street, NW, Washington, DC 20036

## RETURN TO ➤

## CHEMISTRY LIBRARY
100 Hildebrand Hall • 642-3753

| LOAN PERIOD 1 | 2 | 3 |
|---|---|---|
| ▓▓▓▓▓ | **1 MONTH** | |
| 4 | 5 | 6 |

## ALL BOOKS MAY BE RECALLED AFTER 7 DAYS
Renewable by telephone

### DUE AS STAMPED BELOW

| DEC 0 9 | | |
|---|---|---|
| | | |
| | | |
| | | |
| | | |
| | | |
| | | |
| | | |
| | | |
| | | |
| | | |
| | | |

FORM NO. DD5

UNIVERSITY OF CALIFORNIA, BERKELEY
BERKELEY, CA 94720-6000